BOOKS BY JAMES A. CLARK
AND MICHEL T. HALBOUTY

Spindletop
The Last Boom

THE
LAST
BOOM

JAMES A. CLARK &

THE LAST BO

MICHEL T. HALBOUTY

RANDOM HOUSE / NEW YORK

TN
872
T4
C54

Library of Congress Cataloging in Publication Data

Clark, James Anthony, 1907–
The last boom.

Designed by Antonina Krass

1. Petroleum—Texas—History. I. Halbouty, Michel
Thomas, 1909–   joint author.   II. Title.
TN872.T4C54   338.2′7′28209764   72–2019
ISBN 0–394–48232–8

Manufactured in the United States of America

First Edition

TO WILDCATTERS, OLD AND YOUNG,

WHEREVER THEY MAY BE . . .

CONTENTS

BOOK TWO / THE EXPLOITERS

BOOK THREE / THE AFTERMATH

1
THE
BELIEVERS

chapter one

THE PROPHETS

A bit of a charlatan, a bit of a poet, something of a dreamer and always a promoter, Columbus Marion Joiner shambled back and forth across Rusk County, Texas, during the 1920's telling anyone who would listen that the drought-stricken fields and pine-studded hills were floating on an ocean of oil. He was an old wildcatter, bent by his years, by a crippling illness, and by a harsh decade of unrelieved failure, but his hymn of hope was the only song in an area where depression had been the economic constant since the end of World War I.

Only the poor listened. They listened, and they believed because they needed desperately to believe. When he spoke to them in their farmyards they gazed out across their stunted corn and raveling cotton and envisioned derricks sprouting from the parched and cracking earth.

They believed against all reason. In the past, other wildcatters and established oil companies had drilled seventeen dry holes in the county. Scientists for major oil companies had examined every foot of the land with every modern oil-seeking device and almost to a man had condemned it. The few geologists who were not completely convinced that Rusk County was barren of oil were not heeded in the oil company board rooms.

But while the oil companies frowned on Rusk and neighboring counties, speculators in distant cities had caught the postwar oil fever. Discovery of rich fields on the Gulf Coast and in North Central Texas apparently had convinced them that a hole punched anywhere in the Lone Star State would spout oil. Time and time again promoters acquired oil leases on thousands of Rusk County acres to sell to speculators. Almost without exception the purchasers allowed the leases to expire or they grew tired of paying rentals when no successful wells were drilled.

Joiner was an exception. In early 1920 he bought leases on 320 Rusk County acres from an Oklahoma City syndicate. His original plan was to sell the leases at a profit. But first he journeyed down to Rusk County from his home in Ardmore, Oklahoma, to see what he had bought. He arrived as a lease-broker. What he saw made him remain as the wildcatter he truly was. And his promise to tap a "treasure trove all the kings of earth might covet" had by the autumn of 1926 brought him leases on more than 4,000 acres at little or no cost.

The farmers had believed. Now they waited for him to sink his well.

Rusk County lay in Northeast Texas only one county removed from the Louisiana border. It was bounded on the north by Gregg County, on the south by Nacogdoches County, on the west by Smith and Cherokee Counties, and on the east by Shelby and Panola Counties. There was not a major city in the entire seven-county area. Dallas was more than a hundred miles to the west of Rusk County; Shreveport was more than fifty miles to the east; Texarkana was almost seventy-five miles northward. Houston, the state's oil mecca, was a hundred and sixty miles almost due south.

There were only two communities of any size in the county—Overton and Henderson—and they were no more than villages. Neither had a paved street. Henderson, however, was the county seat and boasted a newly built courthouse. Overton was a whistle stop on the Missouri-Pacific main line. Some twenty miles separated the two communities. Between them and about them were farms, pastures and rolling hills covered with pines, sweet gums, oaks, and bois d'arc trees.

A few cotton gins and small sawmills were the only industrial plants in the area. The people in this area had been living off the soil since the county's creation in 1843. Cotton, sweet potatoes and corn were the chief crops, and some cattle and poultry were raised. All of these products had brought good prices from 1914 to 1920. But in the autumn of 1920 farm prices collapsed across the United States, and the entire American economy was in a depression by the spring of 1921. Industry bounced back, but the farmers did not recover. The share of agriculture in the national private production income was 22.9 percent in 1919; it was 12.7 percent in 1928–1929. In Rusk and adjoining counties the situation was made worse by dry seasons that eventually evolved into a full-scale drought. Mortgages and bills went unpaid.

Rusk County was fundamentalist country: when a farmer prayed for rain he went down on his knees to appeal to a personal God. In manners and attitudes the people were more akin to people in the Deep South than to Texans in other sections of the state. Negroes lived just this side of bondage. Most of the whites were of Scotch, Irish and English origin. If they could be generalized, they were physically tough, quick to anger and slow to forgive despite their religiosity. They were warm and generous to their own kind, and friendships lasted a lifetime.

It was not surprising, then, that they opened their hearts as well as their gates to Columbus Joiner. He too was country-bred, and he quoted easily from the Bible he oftimes carried beneath his arm. He did not drink or smoke or use profanity. That he was unusually attractive to women was accepted with sly amusement.

Joiner was born on March 12, 1860, on a farm in Lauderdale County, Alabama. He never knew his father; Corporal James M. Joiner of the Confederate army was killed in 1864 at Jackson, Mississippi. His mother Lucy died when he was eight. An older sister raised him. The family lived in a log house. Joiner cut trees, built fences and picked cotton. At the age of twelve he picked 400 pounds of cotton in a working day and won a twenty-five-cent prize in addition to his daily wage.

He attended school for only seven weeks, but his sister taught him to read, using the only book in the house, the family Bible. He learned to write by copying the Book of Genesis. Several citizens of the nearby community of Center Star were so impressed by the boy's desire for learning that they gave him other books to read . . . and he began dreaming of worlds beyond the log house and the farm. At seventeen he left home and roamed the South, traveling as far as Texas in his wanderings. But at twenty-one he returned to Alabama, got married and managed to open up a dry goods store at Muscle Shoals Canal.

Two years later he was operating a store in Lawrenceburg, Tennessee, and studying law with an attorney who officed above the store. He was a failure as a lawyer, but in 1889, at the age of twenty-nine, he was elected to the lower house of the Tennessee legislature. He served two years, got his fill of politics, and went back to merchandising.

Then in 1897 he moved to Ardmore, Oklahoma, to join his sister, who had married a Choctaw and was living with the tribe. The Choctaws were impressed by Joiner's background and erudition. They employed him to supervise the leasing of tribal lands to white farmers. For the first time in his life, Joiner found himself doing something he enjoyed. By 1906 he owned 12,000 acres of prime farmland and other assets having a total value of $200,000. He had come a long way from the Alabama cotton patch.

The Panic of 1907 wiped him out. He began living a hand-to-mouth existence—buying leases on credit, when he could, with the hope of selling them for cash at a profit. Because land-leasing was the lifeblood of the oil industry, Joiner drifted naturally to the Oklahoma oil fields. And he met a man who was to change the direction of his life—one who said his name was A. D. Lloyd.

Lloyd was born Joseph Idelbert Durham, but he had changed his name to avoid harassment by the women he had left in his wake. In his youth he had worked as a drug clerk while he studied medicine in Cincinnati. Later, bored with being a druggist and physician, he had gone to the Idaho gold rush as a government chemist analyzing ore. It was there that he had become interested in studying the earth. He had become a mining engineer through work in that field and study of books supplied by the United States Bureau of Mines, and through a great deal of self-promotion. He had looked for gold in the Yukon and in Mexico. He had conducted "Dr. Alonzo Durham's Great Medicine Show" across the country, peddling patented medicines he made from oil. He had learned a great deal about geology. And he had learned as much about people and the art of promotion.

Joiner had rounded up leases on a few acres near Earlsboro in Seminole County. Lloyd studied the area and drew up a report on its oil potential, which he said was excellent.

In raising money to drill, Joiner used "Dr. A. D. Lloyd, the nationally known geologist," as his authority.

He drilled, but ran out of money at 3,150 feet. He abandoned the project—and some time later on a nearby lease Empire Gas & Fuel Company drilled 200 feet deeper and opened the great Seminole field.

Joiner drilled near Cement, Oklahoma, again using Lloyd as his authority. He got a "show" of oil and nothing more. Months later Fortuna Oil Company brought in the Cement field.

These terrible misfortunes did not break his spirit. He had become a full-fledged wildcatter. He blithely took credit for discovering the Seminole and Cement fields as he went about trying to raise money to hunt for others.

Most oil leases permitted the wildcatter to hold the acreage for a specified number of years from the date of the lease by paying a nominal rental each year after the first. The wildcatter depended on the sale of part of his acreage—or block, as it was called—for his profit and the expense of drilling his well. Generally, the landowner was to receive one-eighth of the oil found beneath his acres, the wildcatter seven-eighths.

Joiner wildcatted several other areas in Oklahoma without conspicuous success. Between such ventures, he bought and sold oil leases to maintain his family, but he was always hoping to find another area that would satisfy what he by then believed to be his "nose for oil." He found it in Rusk County, Texas.

From 1921 through 1925, Joiner shuttled between his home in Ardmore and Rusk County. Later in 1925 he moved to Dallas; it was nearer to Rusk County and it was full of oilmen and "lease-hounds" who were operating in and around the Texas oil fields. He rented a one-desk, one-chair office in the downtown Praetorian Building. The office rental left him with less than $20.

He was sixty-five years old and in a strange city. An attack of rheumatic fever had left him bent at the waist so that he walked like a man eternally searching the sidewalk for a lost possession. He needed money for his family in Ardmore and for his journeys to Rusk County, but he couldn't "get a deal working." So long before he was ready to do so, he began selling some of his Rusk County leases.

The buyers were not oilmen or professional speculators. They were widows whose names Joiner found in the obituary columns of the Dallas newspapers. He would note the death of a well-to-do man and after allowing for suitable mourning time, would make his call on the widow.

Joiner was a well-built man of medium height when he forced his back to straighten long enough for an introduction or a photograph. His lustrous dark hair was only now beginning to silver. His skin was that of a young boy, a virtue he attributed to a daily ration of raw carrots. His large gray eyes gave his face a Chopinesque sensitivity. And despite his disability, he was vigorous both mentally and physically. Because he had never lost his early hunger for learning, he was well-read in history and literature; he could quote from English and American poetry with the same ease he quoted from the Bible. Though he projected himself as a somewhat shy and humble man, women sensed his basic aggressiveness. They gave him coffee in their sitting rooms and occasionally detained him long after the lease sale had been consummated. In return, he sometimes wrote them long poetic letters, which they frequently misinterpreted.

Joiner sent most of the lease sale money to his family in Ardmore. And in the autumn of 1926, with $45 in his pockets, he set out for Rusk County to build his block to 5,000 acres, the amount he considered minimum for his purpose.

As Joiner labored in southern Rusk County, two men in adjoining Gregg County—men unknown to the old wildcatter and to each other—were preaching with a fervor matching Joiner's that a black bonanza lay buried beneath their feet.

One was J. Malcom Crim, a merchant of Kilgore. The other was a struggling real-estate agent, Barney A. Skipper.

Kilgore, a community of 700, was just a few miles from the dividing county line. While Crim's general store was located in Kilgore, his family farm was across the line in Rusk County—and it was on this acreage Crim hoped to drill a well.

Longview, which had a paved street because State Highway 80 bisected the community, was some twenty miles northeast of Kilgore. Barney Skipper dreamed of oil to the northwest of Longview.

Crim believed oil was under his land because a fortuneteller had told him so. The basis for Skipper's conviction was hardly more credible: his father had told him so since early boyhood. Oil companies were not inclined to act on blind faith and occultism, and particularly when no scientific evidence had been found to support either. So Crim and Skipper preached in vain.

Crim was regarded highly in Kilgore because he was a civic booster who acted as well as talked, and because he extended credit for food, clothing and implements to the despairing farmers in his trade territory. The land he wanted to drill had been acquired by his great-grandfather, Colonel Benjamin Franklin Thompson, in 1844. Colonel Thompson had paid $9,000 for 18,000 acres. At his death the land was divided among his children.

A son, John Martin Thompson, became a wealthy lumberman; he was among the first to exploit the East Texas pine. Before his death he called in his heirs and asked them to help him allocate his wealth between them. His daughter Lou Della had married W. R. Crim, who owned the general store in Kilgore. Her future was secure. Recognizing that her brothers were eager to enter business for themselves, she was satisfied to inherit the family home and the 900 acres on which it sat. The sons divided up their father's money and business properties.

Mrs. Crim bore her husband four sons and a daughter, the eldest being J. Malcom, who eventually took over the management of the family store. Soon after his marriage, Malcolm Crim and his wife journeyed to Mineral Wells in West Texas to visit her relatives. Alone in town one day, he found himself in front of a fortuneteller's tent. Curiosity nudged him inside.

The woman took his fifty cents—and promptly amazed him. "You have a farm," she said. "It is bounded on the north and west by a creek. There is a railroad line running through it, and there is a big house on a hill. There is oil on your farm —and someday you'll discover it. There is oil on other farms bordering yours. You should get that land if you can . . ."

Crim knew, of course, that Rusk and Gregg Counties had been explored, drilled and generally condemned by oilmen. And he was aware that a dry hole had been sunk near Kilgore in 1915 by a major oil company. To the astute, practical businessman he was, this evidence should have been far more persuasive than the fortuneteller's pronouncement. But she *had* described the farm with its creek, the old home and the railroad tracks. And he tossed aside all skepticism when she told him the initials of his neighboring landowner. He left the tent determined to find the oil he now believed was his.

By the summer of 1921 he had obtained leases on more than 20,000 acres surrounding the Lou Della Crim farm. "Move your drilling rig in here and drill anywhere you please, and I'll give you the lease on every foot of it," he told major and independent oil companies alike as the years passed. His offer was rejected time and again. Finally he had to drop the leases.

But again he tried. This time he assembled a lease block of 8,000 acres. "Eight thousand acres on a silver platter with no strings attached," he offered. "All you have to do is drill a well." Again his offer was rejected. Again he had to drop the leases.

Crim didn't despair; he simply made a strategic retreat. He had not been giving enough attention to his store and other interests. He would get his affairs in order. Then he would try again.

As early as 1913—seven years before Columbus Joiner arrived in Rusk County and Malcolm Crim found his fortuneteller—Barney Skipper was trying to interest oil companies in drilling northwest of Longview. Skipper was a superb salesman—he had to be to eke out an existence as a real-estate agent in that time and at that place—but in trying to promote an oil deal he was peddling a product for which there was no market.

He was a stout, broad-faced man, almost puritanical in his habits and attitudes. And he revered the memory of his father, a gentle farmer who was referred to kindly by fellow citizens as Old Man Bill. As a growing boy, Barney Skipper had heard his father say at countless farmers' gatherings that oil was beneath their lands. "It'll be brought up one of these days," he would say. "It'll make you rich." Then he would add: "If we don't get it, maybe my boy will." And he would pat Barney's head. "Won't you, boy?" he would ask. Barney always answered, "Yes, sir."

But wanderlust seized Barney Skipper in his young manhood. Like Columbus Joiner, he wanted to see the far places. Like Joiner, he roamed the South—and he worked in mercantile stores as Joiner had done. Skipper's jut-jawed manli-

ness, his courtesy and, above all, his piety made him a successful salesman. He left a good job in Birmingham because
he was homesick for Texas. In Dallas he became the head
salesman in a clothing store.

Life should have been pleasant for him. He had a good job,
a wife and a baby son. But Old Man Bill's assertion that if no
one else did, his son would find the oil beneath Gregg
County, was to Barney Skipper a promise unkept, a debt
unpaid. On an August morning in 1911 he told the store
manager he was quitting.

"Buy why?" asked the bewildered manager. "I just gave
you a raise—"

"If I can sell for you, I can sell for myself," Skipper replied.

He arrived in Longview with his wife and son and
$307.50. He bought a house—a shack, really—for $300 and
put the remaining $7.50 in the bank. "I'll get us something
to eat," Skipper told his wife. "It's going to be root hog or
die."

He rooted. He walked from village to village, farm to
farm. He couldn't afford a horse and buggy; an automobile
was out of the question. He bought farmland on credit and
sold it for cash—when he could make a deal. And he talked
of oil. People in Longview wondered how he kept his little
family alive. But the Skippers survived. And in 1914, after
World War I broke out and the demand for farm products
soared, he was able to buy an automobile.

During the boom years of the war he sometimes took a
parcel of land as his commission on a deal. To the people
who questioned him about Gregg County's oil potential he
always replied: "Be patient. We'll get the oil. It'll come."

But it didn't come. The boom years passed. The 1920's
brought depression, and still Skipper talked of oil. He added
an insurance agency to his real-estate business to help keep
his family alive.

In 1924 his hopes—and his sales talk—were struck an almost crippling blow. A wildcatter named Patrick White

drilled a dry hole three miles east of Longview. He had not conferred with Skipper. Had he done so, Skipper would have told him that the oil was to the northwest of Longview, as he continued to tell the farmers and oilmen. But the dry hole had a demoralizing effect on Gregg Countians. Some even called Skipper crazy when he continued to preach of oil. "Someday you'll see," Skipper would say.

chapter two

"THE EXPERT"

"Here comes Old Man Joiner, fannin' his butt!"

Walter Tucker looked up from his counter in the Overton general store. Joiner was walking toward the store from the depot, waving a friendly hand at the raucous kids who were scooting down the dusty street and still shouting the identity of the only passenger to alight from the Dallas train. Tucker grinned. Joiner *did* look like he was fanning his butt. Walking as he did, bent forward at the waist, his arms and hands swung freely behind his back.

Tucker got a slab of cheese from a case and cut several slices from it. He opened a box of crackers. Joiner always ordered a dime's worth of cheese; Tucker supplied the crackers free of charge. He placed the food on a piece of butcher paper as Joiner walked in.

"Come in, Mr. Joiner," Tucker said cordially. "How're things going?"

Joiner smiled. "Unbelievable, Walter. Simply unbelievable."

Tucker chuckled. He had never once heard Joiner complain during all the years the wildcatter had been coming to Rusk County. Impulsively, Tucker pushed the cheese and crackers behind a box. "Come home to supper with me," he invited Joiner.

"I'm honored, Walter."

Joiner went home to supper with Tucker that October day, and before the meal was over, he had been invited to stay in the Tucker home and to use a storage room at the store for his office until he was ready to drill his well. There was no mention of payment.

Tucker was a slender, handsome man in his early forties. His wife, Leota, was an attractive, dark-haired woman, hot-tempered but full of sentiment, quick to tongue-lash but quicker to love those close to her. They had three children —John, a strong, slender lad of fifteen, and two older daughters, Mary and Beverly. It was a happy family that welcomed Joiner into the big two-story home.

Tucker could ill afford a star boarder. Business was bad. The store, owned by Tucker and his brother-in-law Wilford Florey, hadn't shown a decent profit in months. Many farmers couldn't pay their bills, and neither Tucker nor Florey could bring themselves to cut off credit to their customers. But Tucker admired Joiner. He had helped him from the beginning of the wildcatter's visits to Rusk County. Tucker knew everyone in his trade territory, and he had opened farm gates for Joiner that the wildcatter could not have opened for himself.

Tucker hoped that Joiner would drill a well and find oil, but he had not yet been convinced that the oil was really there. But the hope—and his own financial condition—now prompted him to help Joiner even more. On previous visits, Joiner had made his way from farm to farm via The Jitney, the name given to a Model-T Ford whose driver was willing to go anywhere for gasoline and a small fee. Now Tucker turned his car over to Joiner. The wildcatter couldn't drive, so John Tucker and his sisters acted as chauffeurs.

Tucker's only assets were his home and his share in the store. He held no property, but Leota owned a 306-acre farm near Kilgore she had inherited. It brought in a small rental. Efforts to sell it attracted no buyers. Tucker's resources dwindled as Joiner continued to build his block.

Early in 1927 Tucker raised enough money for Joiner to travel to Fort Worth. In the lobbies of the Texas and Blackstone Hotels—where oilmen interested in West Texas and Central Texas fields gathered and conducted much of their business—Joiner began a search for his friend of Oklahoma days, A. D. Lloyd.

Lloyd's restless feet and more restless imagination had carried him through a dozen oil fields and countless promotions since his Oklahoma ventures with Joiner. He had become involved in Mexican politics and Mexican business deals; he had formed oil companies and gas companies which bore his name. Still he was little known in the oil industry. Those who knew him told Joiner they had not seen him in months. But a room clerk at the Texas Hotel said he had heard that Lloyd was in Mexico and was expected in Fort Worth soon.

Joiner needed Lloyd badly. He decided to wait. He waited many days before Lloyd's booming voice woke him from a nap he was enjoying in a fat chair in the hotel lobby.

Although Lloyd supposedly had income from several sources, he told Joiner he was broke. But he did have a new car. They set out for Rusk County. The Tuckers were getting another star boarder.

★

Lloyd was six feet tall and weighed 320 pounds. Though he said he was seventy-three—six years older than Joiner— he looked no more than sixty. He was a gray-haired, powerful man with the restless energy of a race horse. Leota Tucker quickly found out that he had an appetite that was never satisfied. Watching Lloyd shovel fried chicken from platter to plate, she would cry out, "My God! He's going to eat the legs off the table!"

Joiner introduced Lloyd to the Tuckers and to other Rusk Countians as "Doctor Lloyd, a great geologist." He did not explain the title and neither did Lloyd. The natives called the big man Doc and let it rest there.

Lloyd went to work. In boots, khakis and a Mexican sombrero, he tramped southwestern Rusk County over terrain his new car could not challenge. He would stay gone for days at a stretch, spending the night where exhaustion halted him. He was studying the area, mapping it and preparing a report on its oil potential.

So gregarious was he, so fond of bootleg whiskey, so bold in his pursuit of unattached females, so captivating when unwinding tales of faraway places, that people flocked to him wherever he alighted. For the children he would bare his head and invite them to lay a finger in a trough in his skull. "A Mexican bandit gave me that when I was fighting against Pancho Villa," he would explain. "Split my head like a watermelon with a four-foot sword!"

Like Joiner, he was a man of considerable learning, and it was revealed in his speech. But while Joiner leaned toward the romantic poets and would spice his sales talks with aphorisms and bits of doggerel of his own creation, Lloyd, when serious, would speak with the apparent authority of a scientific scholar in several fields. The natives gauged him as a man who had studied much, who had the ability to apply what he had learned, but who had too much of the adventurer in him for conventional success.

Because he cut such a colorful and distinctive figure, and

because he implied so much about his past and revealed so little, it was inevitable that stories about him would circulate. They only made him more interesting to the natives. One tale had it that Lloyd had been married six times and had fathered a great number of children. He confirmed this story, one of the few of which he took cognizance. To the farmers, six marriages was such an awesome undertaking it could not be condemned.

But Lloyd's social life did not keep him from his work. By early May he was ready to reduce his studies to writing. And that was the signal for Columbus Joiner to pay a call on Mrs. Daisy Bradford, a comely widow who owned a 975.5-acre farm almost midway between Overton and Henderson—and right in the heart of his block. He had obtained a lease on the farm on August 11, 1925. He had been delinquent in paying the lease rental of fifty cents an acre. It was time to make amends for that.

The first owner of record of the land including and surrounding Mrs. Bradford's farm was a Mexican, Juan Ximines. In the 1830's, when Texas was still a part of Mexico, the government awakened to the fact that East Texas and other remote areas were being colonized almost exclusively by emigrants from the United States. Mexicans were not inclined to leave the warmth and security of San Antonio and other South Texas communities to settle frontier outposts. The government took steps to remedy what it considered a dangerous situation: it urged hundreds of San Antonians, including Juan Ximines, to relocate in East Texas on lands it would provide.

Ximines journeyed northward in 1835 with misgivings tugging stoutly at his elbow. No fool, he was aware that relations between his government and the Anglo settlers had deteriorated and open warfare was imminent. Considerable numbers of Mexican troops were stationed in San Antonio. Who knew what kind of protection to expect in this wild country?

He arrived in Nacogdoches, and on August 12 put in his order for a grant of land. On August 31 he was granted 4,428.5 acres some forty miles to the north of Nacogdoches, in what would become southwestern Rusk County. He paid his tax, about a cent an acre. But without even looking at his land, he sold it for $50 to an Anglo colonizer, Colonel Frost Thorn, and set forth for Louisiana. His profit on the transaction was $4.40.

As Ximines had feared, war came, and Texas emerged from it a republic. And later, in 1840 Colonel Thorn sold 975.5 acres of the Juan Ximines survey to a wealthy newcomer, General Andrew Miller. The price was not revealed, but it is believed it did not exceed $1,000.

General Miller built a spacious home on the land, and his son, Doctor Henry L. Miller, transformed it into a showplace. A verandah circled the house; from it could be seen the flowering fields and the thick stands of pines, willows, sweet gums and oaks. Doctor Miller's children, Daisy, Clarence and Kenneth, were reared in an atmosphere of placid comfort.

As a girl Daisy Miller was beautiful, witty and stubborn. She was courted by many admirers, but she waited until she was thirty before she married, in 1901. Her husband was Dr. William M. Bradford of Henderson. The couple took over the farmhouse as their home, but Dr. Bradford died on January 9, 1904, the third anniversary of the issuance of the couple's marriage license. In the home that had been built for the restless feet of children and the measured, quiet life her husband had envisioned, Daisy Bradford lived alone.

As the years passed, she spent much of her time in Henderson where Kenneth Miller operated the Just Right Hotel, a typical village hostelry. Clarence Miller, Daisy's other brother, delighted in the life of a rural peddler; he traveled

East Texas, calling at farmhouses whose owners seldom got to town. Both men were warm and friendly and devoted to their sister. If she was lonely or embittered by her widowhood, it was not obvious to her acquaintances. "Miss Daisy," as she was called, busied herself in helping others.

Intelligent and vivacious, she was fifty-four when Columbus Joiner entered her life in 1925. He had been quick to recognize that an appeal to greed—the chief weapon in his promotional arsenal—could not be used in this instance. Instead, he had painted a picture of what an oil discovery would mean to the despairing people of the region. He had also spoken of schools, parks and museums. He had been modest in the presentation of his credentials as an oilman, but he had hinted—and had backed it up with quotations from the Scriptures—that he had an ally in the Lord. But he had not been completely the serious oilman. He had drawn on a fund of droll stories to entertain her, but these stories had been designed to call attention to his tenure in the Tennessee legislature and to other occasions when he had received the attention of the citizenry.

Daisy Bradford had been flattered, amused and impressed. She had agreed to lease the farm to Joiner. And when he had proposed to drill his first well on her property if she would only relinquish a fourth of her one-eighth royalty, she had readily accepted.

Now, in May of 1927, Miss Daisy still found Joiner amusing and interesting, but she was a trifle vexed about the delinquent rental payments and the seeming lack of progress. Still, since no one else had approached her about drilling on her land, she accepted his apologies and excuses, and their friendship and business relations continued.

chapter three

THE BIG PITCH

Early-day oilmen drilled for petroleum where there were obvious manifestations of its existence, such as oil seeps or water wells where oil had risen to the surface. Petroleum geology was a comparatively young science—and petroleum geophysics even younger—when oilmen explored Rusk and Gregg Counties during the 1920's. But these scientists had learned that oil migrated through porous strata until it reached what they called structural traps, points at which it could go no farther. The oil accumulated and a pool was formed. Not that all structural traps contained oil—far from it. But oil companies were not inclined to be the first to drill in areas where such evidence did not exist.

The most common structural traps the scientists sought were anticlines, faults, salt domes and "noses." An anticline is an arch with the strata dipping on either side as from the ridge of a house. A fault is a break or dislocation of strata, an interruption in their continuity so that on either side they

are elevated or depressed. A salt dome is actually an anticline, but one of a very particular kind. Salt, geologists have
learned, will move and flow under pressure to points of
weakness, pushing its way up from where it has been deeply
buried and shoving aside overlying layers of rock, leaving
them fractured and tilted. On the Gulf Coast of Texas and
Louisiana, salt domes had caused entrapment of great quantities of oil and so were eagerly hunted. A nose is a semi-
closed structure, sometimes referred to as a monocline.

But geologists and geophysicists had found no structural
traps of consequence in Rusk and Gregg Counties. Indeed,
on several occasions structures were mapped where more
thorough study revealed there were none.

A few geologists, however, were reluctant to dismiss the
area completely because of other features they found interesting and perhaps worthy of study. One of these was a huge
bulge in the earth's surface referred to as the Sabine Uplift.
The bulge centered in northwest Louisiana near the Texas
line. But it was so big—roughly eighty miles long and sixty-
five miles wide—that it extended far into East Texas, being
elongated in a southeasterly direction. It reached down into
eastern Gregg County and on into central Rusk County and
beyond to the south.

Oilmen customarily drilled on or near the top of a structure; they had learned that if oil were present, that's where
they would find it. The farther out on the flanks of a structure one drilled, the less chance there was of finding oil.
Despite its size, the Sabine Uplift was a structure of some
kind; in the past, tests had been drilled on it in Rusk County,
but they had been dry. Since there were no more structures
to be found in the county, there was no sense in drilling
further, oilmen reasoned. They could not afford to cater to
the whims of a few geologists who really couldn't explain
why they were reluctant to dismiss the area.

Then, on March 19, 1927, while Doc Lloyd was still studying southwestern Rusk County and before Joiner had com-

pleted the assembly of his block, Humble Oil and Refining Company brought in an oil field in Anderson and Cherokee Counties, some fifty miles southwest of Joiner's acreage. The field originally was called the Carey Lake field, later re-named the Boggy Creek field. The discovery well was drilled on the flank of a salt dome, and the oil was found in the Woodbine sand.

The Woodbine, as it generally was called, was a formation first recognized near the turn of the century as a prolific water-bearing stratum that supplied water to Dallas and the surrounding areas. As was the custom, it was given the name of the small community, north of Dallas, where it out-cropped at the surface.

But Woodbine became a word on the tongue of every oilman and every oil investor—and even became familiar to many Texas laymen—when in the early 1920's it yielded great quantities of oil from such North Central Texas pools as Mexia, Wortham and Powell. These and other pools were located against geologic faults, and oilmen hunting oil in the Woodbine followed a line of faults that led them from North Central Texas across Northeast Texas and into Arkansas and Louisiana. The route skirted Rusk and Gregg Counties in a great arc whose nearest point was eighty miles away to the northwest.

The discovery of oil in the Woodbine at Carey Lake, how-ever, rekindled Humble's interest in nearby Rusk and Gregg Counties. Teams of geologists and geophysicists were rushed there for rapid reconnaissance. The Rusk County team was led by geologist L. T. (Slim) Barrow, who in later years became a Humble president. The Gregg County team was led by geologist E. A. Wendlandt. Because the oil in the Carey Lake field had been found trapped against an under-ground salt dome, focus of the search was on that particular type of structural trap. Not one salt dome was found in either county. It was the same old story.

But the Humble men had hardly completed their study

when Joiner began mailing out hundreds of copies of a report prepared by Doc Lloyd. Entitled "Geological, Topographical And Petroliferous Survey, Portion of Rusk County, Texas, Made for C. M. Joiner by A. D. Lloyd, Geologist And Petroleum Engineer," the report was accompanied by a map, a confident letter from Lloyd to Joiner on the potential of the acreage, and a letter from Joiner to prospective investors extolling Lloyd's ability as a geologist.

The map clearly indicated a salt dome. It pictured four major anticlines, including one named for Joiner and one for Lloyd. It showed a fault line running through Daisy Bradford's farm.

The report itself* indicated that an exhaustive study had been made of the area surveyed, and it likely would have been impressive to a geologist who had not examined the region. The language was clear and the scientific terminology correct. Lloyd described with authority the anticlines he had found, the faults, the saline dome. He discussed the Yegua and Cook Mountain formations in the area, familiar names to the sophisticated investor. In great detail he explained the correlation between Humble's Carey Lake discovery well and a much earlier dry hole sometimes called the Rucker well.

The report also pointed out that major and large independent companies had leased large blocks of acreage in the area and had paid "unusual prices for same." Seismograph crews, he wrote, had made "thousands of registrations" on Joiner's block, and the "producing oil and gas sands in the field now developed in the region surrounding the Joiner Well yield large gushers . . ." And, he reported confidently, the accumulation of oil and gas in the area "will be of unusual importance."

The report was remarkable for several reasons, but chiefly because every essential statement pointing to Lloyd's opti-

---

*The complete text of the report appears in the Appendix.

mistic conclusion was completely incorrect. There were no anticlines in the area surveyed, no faults, and no saline domes. There was no Yegua formation, no Cook Mountain. There was no correlation between the Rucker well and Humble's Carey Lake discovery; the Rucker well was extremely shallow and had produced nothing. No major oil company had leased large blocks of acreage in the area and of course no one had paid "unusual prices for same." Seismograph crews had not made "thousands of registrations" on and around the Joiner block, which would have indicated structures; they had made no registrations, proof that there were no structures. Even an observation that "the formations dip South" was incorrect: the area formations generally dipped to the west. And there were no fields, no large gushers, in the region surrounding the Joiner lease.

But Joiner was not mailing out the report copies to geologists and oilmen. Like any good promoter, over the years he had accumulated a "sucker list" of persons who occasionally took a flyer on a risky proposition if it were made attractive enough. Lloyd's report was attractive.

So was his accompanying letter. In it Lloyd flatly stated that Joiner would discover one of the largest oil fields in the world! He wrote that Joiner would find the Woodbine at 3,550 feet and was "certain to make a well" in the fabled sand. He ended the letter by advising Joiner to continue furnishing him with cores and cuttings, an implication that Joiner was already drilling.

But on the date of the letter, June 15, 1927, Joiner was not drilling a well, as the letter implied. He was in Dallas trying to raise money to finance drilling, using the report and letter as his sales tools. He had moved his office from the Praetorian Building to the Gulf States Building, and he had hired a secretary.

She was Dea England, an attractive young woman of nineteen. The business school from which she had just graduated sent her to Joiner's office. Her salary—when she got it—was

$15 a week. She was a most competent secretary, and almost immediately she became imbued with Joiner's dream to drill Rusk County. She devoted herself to the fruition of the dream and to Joiner himself.

Joiner formed a syndicate. Out of his holdings of 5,000 acres, some 500 acres were syndicated and offered for sale. For $25 he offered a one-acre interest in the syndicated holdings, with a pro-rata share in the well, which was on an eighty-acre tract he carved out of the Bradford farm. He valued the well at the arbitrary figure of $75,000. Thus, each $25 invested with him entitled the owner to 25/75,000 interest in the well itself plus 1/500 undivided interest in the syndicate.

His certificate stating the proposal, a slip of paper about the size of a No. 10 envelope, was mailed out with the sales pitch. His mailing list contained the names of policemen, postal clerks, railway employees, bankers, merchants and his favorite targets, widows and doctors.

The initial sales push produced little money. Joiner returned to Rusk County to find that Walter Tucker's store had closed its doors. Tucker was broke. His brother-in-law, Wilford Florey, was trying to raise money to start anew. But Tucker wanted no more of the grocery business. He had read Doc Lloyd's letter and now was fully convinced that oil in great quantities was beneath Rusk County. Joiner offered him a quarter interest in the well for past efforts and future services.

As his first act as a wildcatter, Tucker sold half of his quarter interest in the well to a group of businessmen and farmers who managed to raise $900. With most of the money, Tucker bought cured timbers to erect a derrick within sight of Daisy Bradford's farmhouse. He and Joiner then journeyed south to Crockett, in Houston County, and traded leases for drilling equipment.

On the return trip, a truck loaded with equipment fell through a bridge and into a deep creek. The driver escaped through the cab window and swam to shore. It took three days to fish truck and equipment from the creek. After the truck was reloaded and ready to roll, Joiner winked at Tucker and said, "Bad beginning, good ending, Walter."

With the equipment on the drill site, Joiner went to Dallas to check on his sales campaign. He returned with a driller, Tom M. Jones. Jones had studied geology at the University of California and had worked at mining in Colorado before coming to Texas. He had learned to drill in West Texas. He was in Dallas looking for something interesting to do when he met Joiner. Joiner was broke, but he painted a rosy and interesting picture of the future. Jones, who was in his early thirties, was intrigued. He had an automobile. He drove Joiner back to Rusk County.

What he saw when he arrived at the drill site almost drove Jones back to Dallas. The machinery was rusted and almost worn out. The boilers, which were to furnish steam for the engine and pumps, were ancient and mismated. One was an oil-field boiler of 75 horsepower; the other was a cotton-gin boiler with a rated capacity of 90 horsepower. Together they couldn't generate more than 125 pounds of steam pressure, Jones concluded glumly.

He kicked a joint of drill pipe and muttered, "This pipe isn't anything but a streak of rust." Joiner patted his shoulder reassuringly. "You'll get it down, son," the wildcatter said. "We're going to get us the well of the world."

Tucker, Mrs. Tucker, son John and daughter Beverly pitched a big tent on the drill site and moved in. Joiner hired some local men—for $3 a day and a rosy future—to help erect the derrick and set the machinery. Tucker and John worked in the crew. Mrs. Tucker and Beverly did the cooking and washing.

Nothing could have been accomplished without Jones. He was the only person at the drill site with any oil-field knowledge or experience. Yet by late August the 112-foot derrick was up and the machinery ready to operate.

In his spare time, Dennis May, one of the crewmen, produced a little corn whiskey at a still on Johnson Creek. Before Jones set the machinery in motion to "spud in" the well (commence drilling), May passed around a gallon jug of spirits. The crew toasted the well and May, "for good luck," smashed the jug on the rotary table.

But good luck eluded them. Time after time various pieces of the machinery broke down, the ineptness of the crew being responsible for the breakdowns as often as the equipment's antiquity. Jones spent precious hours each day instructing Tucker and the farmhands in the intricacies of drilling. John Tucker was in charge of maintaining the boilers. With the others he felled the trees for boiler fuel. He kept constant check on Johnson Creek, the water supply. And he worked at the forge where the drill bits were heated and hammered to sharpness. He got little rest, for his mother chased him to the creek in the evenings to fish for the crew's supper. At fifteen, John Tucker was playing a man's role.

But finally the crew's inexperience and the condition of the equipment caught up with them. At 1,098 feet, the pipe stuck and could not be budged. Joiner raised enough money to bring an explosives expert down from Shreveport. Putting his explosives inside an inner tube, the expert then lowered them into the well. He ignited the charge, and the earth shook and rumbled. But when the smoke cleared, the pipe was stuck as fast as ever.

In February 1928 Joiner reluctantly abandoned attempts to salvage the well. They had labored for six months and had brought forth nothing.

Abandonment of the well did not discourage Daisy Bradford. She gave Joiner $100 and said, "Go get the rest of what you need. We're going to get us a well."

Joiner went back to Dallas. Although he had not sold out the first syndicate, he formed a second out of another 500 acres of his holdings and mailed out his sales talk. He sold leases outright whenever he could find a buyer. It was April before he decided he had raised enough money to drill again.

By then Tom Jones, the driller, had gone to Venezuela to work for Gulf Oil Corporation. The local men had returned to their farms or were working at odd jobs in the vicinity. Young John Tucker was picking cotton near Henderson. But Walter, Leota and Beverly Tucker were waiting at home for Joiner's return. Joiner located John in the cotton patch and handed him $76. It was the first money he ever had paid the boy.

"I need you, son," the old wildcatter said. "Get back to the rig."

John went to the rig by the way of Overton. He bought a hat, his first suit of clothes, boots and other apparel. With the money left over, he bought fishing gear, the better to pull the catfish and perch from Johnson Creek and into his mother's frying pan.

On April 14, 1928, a new driller, "Uncle Bill" Osborne, spudded in the second well about a hundred feet from the first effort. Joiner had heard about Osborne in Dallas, and he had hired the elderly man by mail. Osborne was from Louisiana. From the first crew, Dennis May and Glenn Pool returned to work. Mrs. Tucker prepared meals from groceries bought on credit from the new store Wilford Florey had opened in Overton.

As before, the work went slowly. There was no generator for lighting, so the crew worked from sunup to nightfall. Some vital piece of equipment seemed always to be in need of repair. The rig would be shut down for weeks at a time while E. J. Teller, the owner of a Henderson machine shop, overhauled it. After one overhaul, Joiner paid Teller with syndicate certificates. Teller was reluctant to accept them, but decided they were better than the promises of payment he had been receiving.

From that time on Joiner used certificates to pay for supplies and services whenever he could, often discounting them as much as 50 and 75 percent. During this period his grown sons, John and Vern Joiner, came to Rusk County from the family home in Ardmore. They too issued certificates as they journeyed around the county helping Joiner with lease problems. So many certificates were issued that storekeepers and customers alike circulated them as money.

Joiner traveled to Dallas frequently to check on his mail campaign and to sell leases. While he was there on an extended visit, a letter addressed to *The Crew, Joiner Well, Henderson, Texas* was delivered to the drill site. The crew gathered around as Mrs. Tucker opened the envelope. Inside was a small beige booklet. Printed on the cover were the words, "Love Letter to Heloise—by C. M. Joiner."

Mrs. Tucker opened the booklet and began reading. Her eyes widened in astonishment, then narrowed in anger. "Why, that old son of a bitch!" she cried. "While we're down here breaking our backs, he's in Dallas writing love letters to some damned woman!" She hurled the booklet into the brush. A crewman retrieved it.

When Mrs. Tucker's anger cooled, she was coaxed into reading the booklet aloud. She began: "My own darling Heloise—my heart warms to every breath of you. I can never feel cast down to earth with the thought of you awakening the sweetest hopes, the highest aspirations . . ."

By now Mrs. Tucker had forgotten her earlier anger. Her compelling voice was the only sound in the small clearing. To her unsophisticated listeners, Joiner's words were purest poetry, and so great was their regard for the printed word that Joiner forevermore would live in their imaginations as a man a cut above mere mortals.

"Oh, what melting delight in your kisses!" Mrs. Tucker read. "If a butterfly could sip nectar from your lips it would be intoxicated and would hide behind a flower, or would fly away into the glorious sunlight while your voice would rise to blend with the voices of the angels . . ."

On and on she read, the dulcet phrases falling from her lips as easily as if she had written them herself. No one stirred; the audience was enthralled.

"I hope these words, though crudely expressed, may fall upon your heart as do the dewdrops that refresh and sweeten every flower. You are the flower of my heart; the master soul."

There was a long moment of silence after Mrs. Tucker closed the booklet. Then she said softly, "I wonder which one she is . . ."

Uncle Bill Osborne snorted, "I don't care. Let's get back to work!"

While the second well was being drilled, two things oc-curred that further damned Rusk County in the eyes of most oilmen. First, the Texas Company (now Texaco) leased some open acreage south of Henderson. With excellent equip-ment, an experienced crew drilled to 3,578 feet in less than two months. Not a drop of oil nor a grain of Woodbine sand was found. Walter R. Smith, the company geologist on the well, visited the Joiner drill site to inform the crew it was wasting its time. He resorted to a dreary oil-field cliché to drive home his point. "I'll drink every barrel of oil you get out of that hole," he said.

Dennis May, the moonlighting moonshiner, was re-strained by other crewmen while Tucker, with a wintry smile, thanked Smith and invited him to leave.

About the time the Texas Company was leasing its acre-age, it was reported that L. W. McNaughton, a Humble geologist, had mapped a structure near the tiny village of London, just east of Overton—and in the direction of the

Joiner acreage. The news stirred some interest, and even more excitement was created when Humble leased 2,083.34 acres and assigned a one-third interest to Gulf. But the excitement promptly died when other Humble geologists made a more thorough study. There was no structure.

McNaughton had made his study near London because Humble was an aggressive company led by daring and resourceful men who gave considerable rein to their field men. Even after the fruitless geological and geophysical studies of Rusk and Gregg Counties following the 1927 discovery of the Carey Lake field, Humble decided to set up an East Texas office in Tyler, a small town in neighboring Smith County. H. J. McLellan was in charge. He was joined by E. A. Wendlandt, the geologist who had helped supervise the 1927 Rusk–Gregg reconnaissance. Working with them was another outstanding geologist, G. M. Knebel.

Though McNaughton's deduction had been incorrect, Wendlandt recommended that Humble keep the acreage it had leased near London. The company readily agreed, chiefly because the leases were obtained so cheaply. This acreage was the first Humble had leased in Rusk or Gregg Counties.

Wendlandt had made the recommendation because of some studies of his own. He had learned that old wells drilled in Smith County—to the west of Rusk County—had found the Woodbine, but that the sand carried salt water, not oil. He also had learned that the old wells drilled on the Sabine Uplift in eastern Rusk County, some thirty miles from the salt water wells, had not found the Woodbine at all. The Woodbine, then, "pinched out" somewhere in that thirty-mile swath, perhaps against the Sabine Uplift. It likely would be bearing salt water all the way, but on the other hand . . .

Meantime, Knebel had mapped a nose in Gregg County south of Longview near the Sabine River. He recommended leasing and drilling the area. Humble landmen went to Barney Skipper, the Longview oil prophet, and employed him to lease a block around Knebel's structure. Skipper needed and appreciated the work, but he did not hesitate to tell the Humble landmen they were making a great mistake. "The oil's to the north and west of Longview," Skipper insisted— as he had been insisting for almost two decades. "That's where you ought to lease and drill."

The landmen thanked him for his advice and told him to lease the Knebel block. Skipper did as he was told. Humble again assigned one-third of the acreage to Gulf. Humble drilled what it termed geological wells to learn the composition of various formations. Whatever they were, no oil was found nor was anything else learned that might have led Humble to immediately drill other wells on the block. But again, on recommendation of its regional geologists, the company maintained the leases.

The obvious interest of Humble and Gulf in the region attracted the attention of other companies at times, but the only oil company representative to visit the Joiner drill site during this period was the Texas Company geologist who told the crew it was wasting its time.

On a March morning in 1929, Uncle Bill Osborne threw up his hands in disgust and walked off the drilling rig floor. Fourteen days earlier the drill pipe had twisted off deep in the bore hole. All efforts to fish the broken segment to the surface had failed.

Osborne had drilled to 2,518 feet, considerably deeper than he had believed the old equipment would drill. It was his opinion, he told Joiner, that fishing attempts had pushed the pipe deep into a wall crevice. "Only a wall-hook will do any good now," he said, but his voice and manner indicated a lack of faith that anything would work.

Joiner seemed not to hear him. He stared at the junky

drilling rig for perhaps three minutes, his face expression-
less. Finally he nodded. "All right, son," he said to the driller,
who was almost as old as himself. He waved a hand at the
others. "You folks go on home. I'll see what I can do." There
was none of the old confidence in his voice.

The next day Walter Tucker went to work as cashier in the
Overton State Bank at $125 a month. Not that he had for-
saken the well—he had too much faith and money and labor
invested in it. He needed the bank job for his family to
survive. But he immediately began working to arouse the
interest of R. A. Motley, the bank president, pointing out
what an oil boom would do for the area at large and the bank
in particular.

Motley was not an easy man to persuade to a point of view.
He had leased some of his acreage to Joiner in 1927 and had
seen nothing result from it. Stone-faced, with a dark mus-
tache and hot, black eyes, there was about him an aura of
mental and physical toughness. It was reported enough
times to deserve a modicum of belief that he sometimes
collected outstanding debts with his fists. But he listened to
Tucker; no man could do less when confronted by such
dedication. And before the month was out Motley had
agreed to finance a Joiner trip to Shreveport to find a wall-
hook. Clarence Miller, Daisy Bradford's salesman-brother,
agreed to drive Joiner to Shreveport. The pair set out on
their journey on April 3, 1929.

chapter four

THE DRILLER

There was no wall-hook to be found in Shreveport supply houses, but the manager of the Pelican Well and Tool Company knew where one was. A driller named Ed Laster had picked up one about a month earlier. He was working on some wells near Waskom, Texas, back across the state line.

"He'll probably give it to you," the manager said, laughing. "Nobody likes to think he'll have to use a wall-hook more than once."

No driller likes to admit he's had to use a wall-hook even once, Miller learned when they located Laster. Fortunately, Miller only said that he had heard that Laster might have a wall-hook.

"There used to be one around here," Laster admitted, but he spent ten minutes kicking around in the weeds before he "found" it.

The wall-hook was a ten-foot length of pipe designed to twist inside and grip the wall of pipe lost in a drill hole so that the lost pipe could be retrieved. Laster threw the rusty device into the back of Miller's car, waving off all offers of payment.

Joiner had not left the car, leaving Miller to deal with Laster. Now Miller introduced the two men, explaining that Joiner was drilling on the Daisy Bradford farm near Henderson. As they shook hands, Laster couldn't help thinking that Joiner, sad-eyed and weary, was the perfect picture of a woebegone wildcatter with a "poor-boy" rig. Laster had seen many of them in his twenty-three years as a roughneck and driller in half a dozen states.

What Joiner saw was a tall, powerfully built man of forty with curly blond hair topping a weathered, handsome face. He had been watching Laster move about "hunting" the wall-hook, and he sensed about the driller a hard, masculine competency. Joiner immediately wanted Laster in his enterprise.

"Come over and visit us," the wildcatter told Laster. "We're going to get a real well, son."

Laster had no intention of visiting, but he said, "I may do that when I finish up here in a week or two."

A week later Laster was out of a job; the hole he had been drilling was a duster. A friend invited him to drive to Jacksonville, Texas, with him. The route ran through Henderson. They stopped there for coffee at Boyd's Cafe, Henderson's social center. Laster asked about Joiner and Clarence Miller and someone brought Miller to the cafe.

Joiner was in Dallas, Miller said, but he insisted that Laster visit the drill site. Laster agreed. He was appalled at what he saw, but he kept his mouth shut. He and his friend drove on to Jacksonville, near Humble's Carey Lake field. The friend, a salesman, called at the Humble field office, and Laster asked the Humble manager about Joiner and his well.

"We don't pay any attention to him," the Humble man said. "In the first place, there's no oil there—and if there were, that bunch couldn't get it."

Laster so far had not seen or heard anything good about the Joiner venture, but several days later Joiner telephoned him from Dallas. Laster was at home in Shreveport. Joiner wanted Laster to meet him in Overton.

Laster was a competent driller. He had a wife and a two-year-old daughter. Drilling jobs were scarce, but he was confident he would find one. He could see nothing but trouble at the Joiner drill site, but after listening to Joiner's low-keyed but insistent sales talk he agreed to meet the wildcatter in Overton.

He told his wife about the telephone conversation. He grinned ruefully. "He can charm the birds out of the trees, honey."

Walter Tucker drove Joiner and Laster to the drill site. Laster made a more thorough examination of the site and the equipment. He could see that someone had used the wall-hook without success. It was obvious to him that the hole could not be saved, that it would be necessary to move the rig and spud in a new well. He finally decided that with luck and careful handling, the old machinery could be made to function. It was at this point that Tucker told him that the drill had hit gas just below 1,400 feet on the second well.

"We used up a box of matches burning bubbles on the slush pit," Tucker said.

"That's interesting," Laster said.

"Come in with me, son," Joiner pleaded. "I'll give you ten dollars a day—six in cash and four in leases. And when we hit that ocean of oil, you'll be rich right along with the rest of us."

Laster shook his head slowly.

"Come on, son," Joiner begged. "You can get that hole cleaned out and drill right on to pay sand. I know that you can do it!"

Laster held up a hand. "Wait a minute! You can't bring that well in. It's junk from top to bottom. And you can't possibly avoid water, fresh or salt. It's exposed—"

"No!" Joiner cried. "I can't start over! I've spent too much money on this well! People are depending on me, son! All these people around here!"

Laster shook his head. "Nobody'll ever drill that hole, Mr. Joiner. You can forget it."

Joiner argued, Laster argued back and the wrangling ended with Laster asking to be driven back to Overton and his car. There was little conversation as Walter Tucker drove them to town. Laster said goodbye and drove home to Shreveport.

A week later he got a telegram from Joiner. It said: Will do it your way STOP Come at once."

With the help of Dennis May, Glenn Pool and Daisy Bradford's tractor, on May 8, 1929, Ed Laster began skidding the Joiner rig to a new location. About 375 feet from the second well site a sill supporting the derrick caught on a rock and snapped in two. The site Joiner had selected for the Daisy Bradford Number 3 was still more than 100 feet away.

"Drive over to the sawmill and get us another sill," Laster told Dennis May. "We'll pull a leg off the derrick if we try to move without one."

"Not unless you've got ten bucks," May said. "The sawmill's cut off Mr. Joiner's credit just like everyone else has."

Laster had a temper. He kicked a tractor tire and said a few choice words about any driller stupid enough to get caught in such a situation. He calmed down when he saw May and Pool grinning at him. "Oh, hell," Laster said, grinning back, "let's get the damned thing leveled off and drill right here."

To Laster, this was the first indication of just how poor a poor-boy operation Joiner was conducting. His decision to "drill right here" was revelatory of his character. He knew Joiner was in Dallas trying to raise money. He didn't know when Joiner would return. He reasoned that if oil were beneath the sites of the abandoned wells, it was beneath the ground on which he stood. He was a man of action and quick decision. Thus the site of the Daisy Bradford Number 3 was decided by the lack of a $10 sill.

Laster had been impressed by his reception. Joiner had arranged an airy room for him in the Bradford farmhouse. The house was occupied at the time by a schoolteacher, Mrs. Lena Hunt, whose husband farmed the land. Mrs. Bradford also stayed in the farmhouse frequently, and she would visit the drill site almost every day. By now she had become familiar with procedures at the well. Laster had added two local men to his crew and drilling was progressing smoothly. Mrs. Bradford quickly gained an admiration for Laster's ability and qualities of leadership, and she developed a liking for the man himself.

Two days after Laster spudded in the Daisy Bradford Number 3, he had drilled to below 1,200 feet—a feat so remarkable in contrast to the drilling on the abandoned wells that the weekly newspaper in Henderson took note of it in a three-paragraph story.

Two days later Laster ran out of fuel. To his dismay, he learned that the crews on the abandoned wells had cut and used all of the available firewood on the Bradford farm. Finally he located a Negro wood-hauler, Dan Tanner, and asked him to supply the boilers. Tanner knew about the Joiner efforts; he had little confidence in being paid. Laster was honest with him.

"I'll do my best to get your money right along," he told Tanner. "If I can't get it that way, I'll pay you out of my share of the oil if Mr. Joiner doesn't—but I'm sure he will."

Tanner agreed to help, but he had no money to buy the seasoned firewood that the boilers required. He brought to the drill site the only fuel he could scrounge—green wood that produced more smoke than fire and heat.

Laster would help the boiler fireman raise steam to power

the machinery. When the gauge showed that there was as much as 115 pounds of pressure, he would hurry to the rig and begin drilling. In ten or fifteen minutes the steam would be exhausted, and drilling could not be resumed until the pressure had been built up again. It was aggravating to a man of Laster's temperament. He controlled his temper at the expense of his stomach. He began to belch his food; fiery stomach pains bent him double on occasion.

Three times the rusty, rotten drill pipe twisted off. Each time Laster fished it out and resumed drilling. Once, while he was lifting the pipe from the hole to change a bit, a chain broke. The pipe dropped back into the hole and was bent. As delicately as a seamstress plying her needle, Laster drew the pipe from the hole, repaired the chain and in two hours had resumed his intermittent drilling.

His first paycheck, mailed from Dallas by Joiner, was not honored at the bank in Henderson. His crew was not surprised; many of their checks had not been honored. Laster drove to Overton to see Walter Tucker. Motley, the bank president, cashed Laster's check and the checks of the crew.

The next paychecks did not arrive on time—and that became the pattern. More frequently than not, Laster drilled with only one or two men helping him. A farmer would drop by and work until noon, then go back to his plowing. Glenn Pool and Dennis May were his most constant workers, but even they would have to leave the drill site to find odd jobs to support themselves and their families.

It was in this way that Laster managed to get the well below 1,400 feet, where gas had been reported in the second well. He found no gas. He continued drilling.

Late in August, as Laster and a farmer, Jim Lambert, were building up steam, a mud ring blew off the ancient oil-field boiler and scalding steam burned them both seriously. They were taken to the Just Right Hotel in Henderson, where Mrs. Bradford and her sister-in-law, Mrs. Ora Lee Miller, nursed them for almost a month.

Joiner came from Dallas. He tried his best to cheer up Laster and Lambert, but he admitted that the financial picture was growing gloomier by the day. "But we'll get her down, boys," he said. "Nothing can keep us from that oil." He left them to return to Dallas to "see what I can do."

As far as Laster could determine, he did very little. When Laster was able to return to work, he could find no workers willing to join him. Motley had cashed Joiner's checks for several of them—checks which Tucker said were no good—but the money had long since been spent. Laster and E. J. Teller, the machinist, repaired the boiler, Teller accepting Laster's word that he would be paid. But the machinery still sat idle.

Now even Laster was ready to surrender.

Rusk County was not the world, nor even the state, although it often appeared to be so to those close to the Joiner project. Other oil fields were being discovered, most of them in Texas, and on October 13, 1929, the Pure Oil Company drilled the discovery well of the Van field in Van Zandt County, two counties and more than sixty miles to the west of the Joiner drill site.

Van was a good field, though not a big one, but it did not produce a boom, as almost every other Texas oil field had done. All of the prime acreage was held by Pure, Humble, Sun, Texas and Shell; there was no room for the independent oilman. Also, and more importantly, the five companies agreed to produce the field on a unitized basis: that is, the companies forsook the customary practice of each drilling and producing as rapidly as possible in order to drain the pool before the others did, but instead allocated the oil and spaced the wells on the basis of acreage owned in the field. Some said the unitization agreement, remarkably advanced for its time, enabled Van to escape the vice, disorder and lawlessness that generally accompanied an oil-field boom. Few dry holes were drilled and the reservoir pressure of the pool was not carelessly dissipated by unnecessary drilling.

Van produced little excitement in Rusk and Gregg Counties, but it did spur Joiner to action. He got in touch with Laster, who was at home in Shreveport, and induced him to return to work. Laster was amazed that after all the dreary months Joiner still had the capacity to renew the natives' faith in the well and in himself. Joiner made a whirlwind tour of the region. He organized a new syndicate, the third one; again the certificates—sold at great discounts—began circulating as money throughout the territory. He went to Dallas and peddled some leases. And he paid several discreet visits to a pair of maiden sisters in a village near Overton. The money he received from them went to pay Dan Tanner, the wood-hauler, and E. J. Teller, the machinist.

Though Joiner was a modest man, he confessed to a small group in the Overton State Bank that he had a way with women. "Every woman has a certain place on her neck, and when I touch it they automatically start writing me a check," the old wildcatter said. "I may be the only man on earth who knows just how to locate that spot." He grinned and added wryly, "Of course, the checks are not always good."

Laster was pleased that he had not been forgotten. The people treated him as an old friend, a comrade in arms against the drought and against the Great Depression that by now had engulfed the entire country. His wife Eugenia and their daughter Caroline came to Rusk County for extended visits, staying at the farm or in a Henderson rooming house. Miss Daisy poured out her love on the beautiful child.

In this euphoric atmosphere, Laster fought the balky machinery and the green firewood and the rotten drill pipe to sink the bit to below 2,600 feet by late March. Then everything went awry at once. Money did not arrive from Dallas, bills and wages went unpaid. Dan Tanner fell ill. Laster temporarily solved the fuel problem by burning discarded automobile and truck tires in the boilers. Still, he could drill only minutes at a time before his steam was exhausted. And then the pipe twisted off at 2,640 feet. It took all of his skill, all of his courage and all the help he could muster to clear the hole of broken pipe. Then his help evaporated and his own core of strength seemed to lose its hardy resiliency.

From the beginning it had seemed miraculous to Laster that the hole had "stood up" as well as it had. Despite the various troubles it had not caved in, to be lost forever. Every working morning he had found the pipe stuck in the hole, but a few minutes of "putting a strain on it" and working it up and down would loosen it. Drilling could then be resumed.

Now he needed another kind of miracle—and one occurred. It was a phone call from Joiner in Dallas, relayed to Laster by Daisy Bradford. Joiner would be bringing some prospective investors to the well site the following Sunday. Please, the old wildcatter begged, be "taking a core" when the party arrived.

A driller on a wildcat well would take a core only when he thought he had drilled to a formation that might contain shows of oil. He would send a hollow tool with a specially designed cutting bit (known as a core barrel assembly) to the bottom of the hole. As the bit rotated in the formation, a sample of the stratum would pack into the tool; a "seat" on the tool assembly would close as the assembly was lifted from the bottom; after the core barrel assembly was removed from the hole, the contents of the tool would be examined.

Laster had no core barrel. He went to Tucker and Motley. They gave him money to rent one in Shreveport. Laster was ready when Joiner and his prospects arrived at the drill site. With Dennis May, Glenn Pool and two farmers helping him, he pulled a core with professional efficiency right before the prospects' eyes.

Joiner—no geologist—examined the core, gravely mutter-
ing as he punched at the mud. "It seems to me that you're
nearing the Austin Chalk, Ed," he said crisply. "The Woodbine can't be much farther down, according to our geologist."

"That's right, sir," Laster said. "I expect to hit the chalk
any day now."

"Fine," Joiner said. "Keep right on with the good work."

And he whisked the prospects away.

Laster sighed as the party disappeared. How long, he wondered, would it take the old wildcatter to part the prospects
from some of their cash . . .

Neither Joiner nor Laster were aware that during this first
year of the Depression two oil companies had made moves
into the general area. First, without fanfare, Sinclair Oil
Company had leased a block near Troup, a village southwest
of Overton in Smith County. No equipment was trucked in,
no drilling crews arrived. Apparently the company simply
intended to sit on its investment.

Meanwhile, the Herbert Oil Company of Fort Worth,
headed by John W. Herbert III, a flamboyant Pennsylvanian
who flew his own plane from oil patch to oil patch, acquired
leases on more than 3,000 acres adjoining the Humble-Gulf
Rusk County block near London. To fellow oilmen who
questioned the transaction, Herbert and his partner Ralph
Weaver explained that their action had been prompted by
an article written by E. A. Wendlandt and G. M. Knebel, the
Humble geologists who had never lost interest in Rusk and
Gregg Counties.

The article appeared in the October 1929 issue of the
AAPG Bulletin, a publication of the scientifically prestigious
American Association of Petroleum Geologists. It was a re-
port on the Lower Claiborne Series of East Texas, a calm

appraisal of and an introduction to new terminology for East Texas generally. It contained little or no information which would prompt a mad rush to drill Rusk County. Indeed, it had not prompted Humble to drill, and Humble would not have allowed its scientists to publish the paper had the company thought they were giving away valuable secrets. It seemed more likely that Herbert and Weaver had leased the block because they believed Humble would not have leased —and kept—its block unless there were strong indications of oil present.

In any event, in April 1930—six months after the article's publication—Herbert and Weaver proposed that their company, Humble and Gulf conduct a joint program of exploration and development. The agreement was signed on June 2. Among other things, it specified that within seven months Herbert would begin drilling at a location selected by Humble. The well would be drilled to 3,500 feet or to sufficient depth to reach the Woodbine, whichever was less.

It appeared that Herbert–Humble–Gulf might reach Columbus Joiner's "ocean of oil" before Ed Laster—providing the Woodbine was present . . . and was carrying petroleum, not salt water.

The spring of 1930 was also a busy period for Barney Skipper, the Longview oil prophet. Skipper was employed by Amerada Petroleum Company to acquire some leases in eastern Smith County. Again he told his employers they were leasing in the wrong area, that in this case the leases were too far to the west. Again he was ignored—and Amerada drilled a dry hole.

Skipper then prepared himself for a supreme effort. He had the telephone company send him directories from Houston, Dallas, Fort Worth, Shreveport, Tulsa and Oklahoma City. From the directories he took the names of major

oil companies, independent oil companies, drilling contractors, oil operators—names of any companies or individuals
who seemed even remotely concerned with drilling for oil.

Then night after night he and his wife and son sat up and
wrote letters, offering to share the Skipper leases with anyone willing to drill them. *They wrote and mailed more than
750 letters.*

The replies—more than Skipper had expected—were not
slow in coming. All were refusals. Not a single one offered
him any encouragement. The wick of his confidence was
almost extinguished. For the first time in his adult life the
hard-charging optimist felt his enthusiasm drain away.

It was revived by a telephone call that came late of an
April evening. The caller was Walter W. Lechner. He was
interested in Skipper's leases. Could Skipper meet him at
nine o'clock the next morning in Longview's Gregg Hotel?
Skipper could and would. He did not ask for credentials.

Lechner had the credentials. He was an experienced oilman, a ready gambler with a deserved reputation for solid
honesty. He had learned about Skipper from an old friend,
Dr. W. D. Northcutt. Lechner was on a business trip and had
stopped in Longview to renew briefly his friendship with
the physician. He told Northcutt he had been drilling a well
in Scurry County, West Texas.

"Why don't you drill one around here?" Northcutt asked.

"I will if I can get enough acreage to warrant a deep hole,"
Lechner said. "Do you think that can be managed?"

"It certainly can," Northcutt replied, and he told Lechner
that Barney Skipper was the man to see. "He's got acreage
from Longview west to Gladewater—and he's been preaching oil since the day he showed up around here."

Lechner was disappointed when Skipper showed him a
map of his leaseholdings. The leases were not in a block but
were scattered over a large area. And they totaled only 2,300
acres.

"We'll have to have ten thousand acres before we can get

someone to drill," Lechner told Skipper, "and they'll have to be in a block."

"Well, the farmers are giving me the acreage," Skipper said.

"Good," said Lechner. "Let's get to work."

Lechner hired a notary public to accompany him as he traveled around Gregg County to obtain leases. Several times he had to journey to distant states to clear titles to properties he wanted. As the block grew, he began trying to interest major oil companies in the enterprise. Despite Lechner's reputation, he was no more successful than Skipper had been. Only one geologist, Robert Whitehead of Atlantic Refining Company, became enthusiastic about the acreage. But Whitehead's glowing report created no interest in Atlantic's main office in Philadelphia.

"I'm terribly sorry my people can't see this as I do," Whitehead told Lechner. "But you've got the world sewed up here, Walter, and don't you let it get away from you."

Lechner didn't intend to.

chapter five

THE WOODBINE

The first professional oilman ever to visit a Joiner drill site on the Daisy Bradford farm paid Ed Laster a call in mid-April of 1930. He was Donald M. Reese, a young scout employed by Sinclair. He had set up headquarters in Tyler, as had scouts for most of the major oil companies, soon after the discovery of the Van field in October 1929. Tyler, county seat of Smith County, was centrally located for scouting all of East Texas, and it was large enough to provide the essential social amenities for the comparatively sophisticated oilmen. It will be recalled that Humble established a field office there after the discovery of Carey Lake field in Cherokee County and the 1927 reconnaissance of Rusk and Gregg Counties. A few more companies followed suit after the Van field came in.

Laster was drilling the Daisy Bradford 3 about a mile from a paved road that ran between Henderson and Tyler. A crooked-letter sign, nailed to a tree beside the paved road and pointing to a dirt road through the woods, announced to the world: C.M. JOINER, BRADFORD 3. Reese negotiated the dusty road and introduced himself to Laster. It was evening, and Laster was alone. He also was hungry for the kind of oil-field talk young Reese could provide. They talked of oilmen they had known and fields they had seen; Reese brought Laster up to date on happenings outside his narrow world.

Reese was only twenty-five, but he was experienced enough to assess quickly the junky drilling equipment and the weary derrick. He also was sensitive enough to understand, with compassion, the ordeal Laster had been undergoing. He was candid with the driller.

"This is the only test well between that Texas Company dry hole south of Henderson and the Amerada dry hole in Smith County . . . That's why I'm here," Reese said. "You've got the only hole drilling for forty miles around or maybe more. And it doesn't look like you're doing much drilling . . . or expect to."

Reese plainly was implying that the Joiner venture was purely a promotional one, that Joiner chiefly was interested in acquiring leases cheaply and selling them at a profit, that the drilling was a carnival come-on.

Laster surprised him. "I'm going to get oil here," Laster said. "I've picked up a few leases for myself."

Joiner had been slow in tendering Laster the leases promised as part of his wages. Aware of this, Walter Tucker, R. A. Motley and Clarence Miller had helped him acquire some.

Laster also had developed a proprietary attitude about the Daisy Bradford 3. Because Joiner had left him to his own devices so much, because he had poured so much of his sweat and spirit into the drilling, because the natives looked to him to fulfill Joiner's promise of glory days, the hole he had bored into the stubborn earth had become *his* well, and he often referred to it as such.

Laster had taken the drilling job with all of its uncertainties as a gamble for oil. Now, with all of the effort he had put into the job, thinking of the well as his own, he believed strongly that he would strike oil if Joiner would bestir himself more vigorously.

Reese grasped these points even though Laster did not elucidate them. Reese still believed the venture was an out-and-out promotion from Joiner's point of view, but he thought that Laster might very well drill deep enough for the cuttings from the hole to be worthy of study by Sinclair's geologists. He asked Laster to save sample cuttings for him. Laster readily agreed, something a driller would not have done under normal circumstances.

Reese began coming to the drill site several times a week, and shortly scouts from other companies began making professional visits. Laster's most regular visitors were Reese, Henry Conway of Amerada, Calvin McMahan of Transcontinental, Jess Hamilton of Sun, Hillary Hebert of Gulf, and Boots Fulton of Shell. Reese and Conway, however, visited the site more often than the others; the others were willing to share in any information Reese and Conway would provide while they scouted what they considered greener pastures.

Laster was stimulated by this attention. At this point he and Joiner decided to create some "lease action" in Dallas. Laster would go into Henderson and telegraph Joiner that he was "taking cores." Telegram in hand, Joiner would hurry to the lobbies of the Baker and Adolphus hotels in Dallas, where the oil fraternity gathered. He would arouse interest with the telegram, then begin his sales pitch about his "ocean of oil."

In this way enough money would be raised for Laster to drill a little longer, a little deeper. But the tactic almost cost young Donald Reese his job with Sinclair. It will be recalled that John Herbert III and Ralph Weaver held a block near London adjoining Humble–Gulf acreage, and had made a deal with Humble–Gulf for a joint exploration effort. No well had been spudded in, however, and there was no indication that one would ever be. It also will be recalled that Sinclair had leased a block near the village of Troup, but had made no effort to drill. But now, with Laster's drill probing deeper into the earth, both Sinclair and Herbert–Weaver wanted up-to-the-minute reports on the Daisy Bradford 3.

Sinclair had no laboratory facilities in Tyler, so Reese was taking his cuttings to the Prairie Oil and Gas Company office where they were examined by Lee Smith, the company paleontologist. Smith reported to Reese that Laster's bit had not yet cut the Austin Chalk, a formation that had to be penetrated before the Woodbine could be reached at its estimated depth. Reese would send this information to his superior in Dallas, Arch Lowrance.

Lowrance, however, was in almost daily contact with Ralph Weaver—and Weaver was hearing regularly from Joiner that Laster was "taking cores," an indication that Laster's drill was in or near the Woodbine . . . or the depth where the Woodbine might be expected to be pres-ent. Lowrance and Smith, the Prairie paleontologist, began questioning the validity of the samples Laster was saving for Reese, an implication that Reese was being "suckered" by Laster.

Reese was confident that Laster had been giving him legitimate cuttings, but now he began going to the drill site at night to steal cuttings from the trough. These cuttings checked with the ones Laster was giving him. Reese tele-phoned this information to Lowrance in Dallas, also point-ing out that the rig had been shut down for lack of fuel on some occasions when Joiner was boasting of "taking cores." Lowrance was not completely satisfied.

In mid-July, Laster did take a core, but he did it more to satisfy his own curiosity than for any other reason. He did it at the suggestion of Henry Conway, the Amerada scout. "You're killing yourself out here, Ed," Conway told Laster. "Why don't you take a core and let me send it over to the Pure office at the Van field? They'll tell you whether you're going to make a well or not. You keep on hanging around here and this old derrick will fall on you."

Laster took a core at 3,456 feet. Conway sent a portion of it to Elmer Rice, the Pure paleontologist at Van field. But he and Laster went to the Prairie office in Tyler and asked Smith, the paleontologist, to examine the remainder. Smith ground up the core and examined it with his instruments. He invited Laster to look at the core through a microscope. Laughing, Laster took a look, but told Smith, "I don't know anything about these bugs you're talking about."

"Right now you're in the Lower Taylor series, and you've got two more formations to cut before you arrive at the Woodbine, if you find it," Smith said. "*If* you find it, it will be at approximately fifty-one hundred feet and carrying salt water."

Laster was crushed by this professional pronouncement. It was generally accepted that the Woodbine—if it were present at all in the area—would be found at about 3,500 feet. So the Humble geologists believed, and Doc Lloyd had confidently written Joiner that the Woodbine would be found at 3,550 feet and laden with oil. Laster had taken the core at 3,456 feet, almost 1,000 feet deeper than his rig—when new—was designed to drill. He had no hope of drilling to 5,100 feet.

Laster had never placed any confidence in Doc Lloyd's report or in his letter to Joiner. He had never met Lloyd; the big man showed up occasionally while the two abandoned wells were being drilled, but at this time he had not visited the Bradford farm in more than two years. Like most other oilmen, Laster felt sure that the two documents were sales tools for Joiner, that Joiner had quoted Lloyd because he had had no one else to quote. Laster's growing belief that oil was beneath his rig had come from his driller's instinct and from his proprietary attitude about the well. In this terrible moment, the be-

lief sustained him. He returned to the rig determined to drill
to at least 3,500 feet; he would not be satisfied without doing
that much.

Laster also was aware that Joiner's lease on the Bradford
acres expired on August 11—less than a month away. He
would have to hurry. Sharing Laster's disappointment to
some extent, Conway assumed that Laster would shut down
the rig. He told the other scouts in Tyler what had occurred.
The scouts reported the news to their superiors. But Lowrance, the Sinclair official, told his scout, young Reese, to
keep tabs on the well "just in case," even though the paleontologist's conclusion seemingly verified the negative information Reese had been sending Lowrance.

Meanwhile, Laster resumed drilling with the help of Dennis May and Glenn Pool. After the third day of drilling, the
weary Laster noticed that his fish-tailed drill bit was "hack-
sawed," indicating that it had cut into hard sand. Around the
well's surface pipe Laster kept two large buckets for cuttings. He examined the cuttings from one bucket and found
several pieces of hard, crystallized sand. His heart began
beating a little faster because the sand contained streaks of
color.

Though it was growing dark, Laster had his men stay with
him to take a core. When the core barrel was removed from
the hole there was nothing on the outside of it to indicate
it had reached oil sand—and Laster quickly decided not to
examine the barrel's contents. But as the men washed off the
rig floor, he noticed bubbles rising in the mud that had clung
to the outside of the barrel. Gas, he told himself.

He tossed the barrel and bits into the back of the car,
telling May and Glenn that he would stay in town the next
day and have the bits sharpened. He drove home to Henderson. He didn't want to be seen examining the core barrel,
so he waited out the night before opening the device at
dawn behind the rooming-house garage. What he found was
about nine inches of the prettiest oil sand he had ever seen.
He had hit the Woodbine!

Almost simultaneously with this magnificent discovery came an alarming recollection: He had left one of the buckets, full of cuttings, on the rig floor!

He drove to the rig at such great speed that he blew the head gasket on his old Nash. He raced on foot across the drill site clearing. He clambered up on the rig. The bucket was gone!

As the hours passed Laster sat alone at the drill site, agonizing and sweating in the July heat. He had kept the discovery secret from his crewmen. He had wanted to keep it secret from all but Joiner in order to allow the wildcatter time to regain his sold leases, if possible, and to acquire additional ones at the prevailing low cost. But now the possessor of the bucket of oil-streaked sand cuttings was in a position to benefit himself or to shout the news to the world. The thief must have recognized the value of the cuttings or he would not have taken them.

Adding to Laster's bitterness was the galling memory of selling one of his own leases a week earlier to get enough money to pay his current expenses. That lease had increased in value a hundredfold—perhaps a thousandfold—because his bit had found the Woodbine.

He was jarred from his sour revery by the sound of an approaching automobile. The driver was Henry Conway, the Amerada scout. Had Conway taken the bucket? Was he here now to say that he had told Amerada of its contents?

Conway walked to the rig waving a telegram. He looked at Laster with sympathetic eyes as he handed the driller the yellow slip. Laster read the telegram. It was from Elmer Rice, the Pure paleontologist at Van field. It said almost the same thing that Lee Smith, the Prairie paleontologist, had told Laster in Tyler—if he found the Woodbine it would be at about 5,100 feet and would carry salt water.

So Conway hadn't taken the bucket.

Laster thanked the gentlemanly Conway and the scout departed. Still haunted by the question of who could have taken the bucket, Laster picked up around the drill site, keeping busy almost mechanically. Forgotten was the blown head gasket on the Nash. He was trying to think of the best way to present the good news and the bad to Joiner.

He was interrupted again, this time by Reese, the young Sinclair scout. Reese walked from his car with a grin on his face. "Well, I found your bait, old buddy," he told Laster.

"What bait?"

"Don't act so innocent. I mean that oil sand in the cuttings bucket. I know you got it from the Van field. You can buy all you want from a roughneck for three or four bucks."

Laster's stomach was churning. He hoped his face didn't show the relief he felt. As carelessly as he could, he said, "If that's what you want to think, fine."

Chuckling, Reese slapped Laster's shoulder. "Salting the well so you can stir up interest," he chided. "You ought to be ashamed."

Laster laughed with him.

Back in town, Laster telephoned Joiner in Dallas and gave him the news. If Joiner was elated, he managed to disguise it. He asked Laster to ship him a piece of the core by bus. Then he said that Doc Lloyd would be coming to the well site to take charge, and that he himself would be in Henderson in a day or two.

Laster was stunned and cut to the heart. After all the work he had done, he felt that now he was being shunted aside. Someone else would complete the well—if it could be completed. He had done more for Joiner than drill the well under terribly adverse conditions. Because the farmers and townsmen liked and respected him, he had been able to induce several to renew lease deals which otherwise would have expired and been lost to the old wildcatter.

That night he met with representatives of the Mid-Kansas Oil and Gas Company and made a deal to deliver to the

company a piece of the core and other pertinent information about the well.

It was not to Donald Reese's discredit that he had believed
Laster had salted the well. For several months he had been
harassed by his superior, Arch Lowrance, who had believed
—or had pretended to believe—the Laster–Joiner lies that
Laster was taking cores when he in fact was not. Lowrance,
however, had believed the erroneous reports of the two
respected paleontologists, and so had Reese. But Reese was
a thorough scout. After he had found the bucket of cuttings
at the rig he had gone to the courthouse in Henderson to see
if Laster, Joiner, or any of their associates had been quietly
buying leases. The records showed the contrary, and the
clinching piece of evidence was the record of the lease sale
that Laster had made within the week. A man sitting on top
of an oil field does not sell his leases.

All professional interest in the Daisy Bradford 3 died
away, and Joiner, in Dallas, did not spread the word of Laster's strike for reasons of his own. But on July 28, 1930—a
Saturday—Reese and Calvin McMahan, the Transcontinental scout, drove to Shreveport to check on the progress of a
deep gas well. On the return trip, Reese suggested they stop
by the Daisy Bradford 3 since it was only a mile off the
highway. As their car moved into the drill site clearing,
Reese slammed on the brakes. "Well, I'll be damned!" he
muttered.

Standing around the rig and gathered on the rig floor
were Joiner and his son John, Doc Lloyd, Ed and Eugenia
Laster, Walter and Leota Tucker, Daisy Bradford and Clarence Miller, and J. H. Shelton, a Dallas automobile dealer
who had invested fairly heavily in the well. Laster, Lloyd
and John Joiner had just retrieved a core barrel from the
well. The three men grabbed the barrel and ran with it to

John Joiner's Ford. John Joiner drove up a dirt road into the woods. Reese and McMahan followed. John Joiner parked under a shade tree and Reese stopped his car some yards away. But John Joiner, Lloyd and Laster simply sat in the Ford; they were willing to outwait Reese and McMahan.

Reese managed to turn his car around. He drove back to the well. He got out of the car and approached Columbus Joiner and the others. He was met by hostile looks from all but the old wildcatter. "I've got to have a piece of that core, Mr. Joiner," Reese said. "I'm already in trouble because of this well."

Wiser than his associates, Joiner nodded his head. "Come with me, son," he told Reese, and he led the young man back to the car where McMahan was waiting. Joiner stuck his hand inside the car and placed a piece of ashy sand about the size of a thumbnail in McMahan's hand. "Don't look at it here," the old wildcatter cautioned.

Reese and McMahan drove away from the well. At the highway they examined the piece of core. It reeked of crude oil. They drove down the highway a bit to lay in wait for John Joiner's Ford, hoping to get a larger piece of the core. But during the night their quarry eluded them.

The following Monday Reese reported what had occurred to Arch Lowrance in Dallas. Lowrance told him to have Lee Smith, the Prairie paleontologist, examine the bit of core. Smith stuck to his conviction that Laster had never penetrated the Austin Chalk and was at least 1,500 feet above the Woodbine. His conclusion was that the core was salted, a not-uncommon practice.

Reese insisted to Smith and to Lowrance that he believed strongly that the core was genuine. They paid him no heed. He carried the core around in a penny matchbox to show the other scouts. They laughed at Reese without malice. They kidded that Reese and McMahan had made up the entire story. They called the pair "Joiner Reese" and "Joiner McMahan," implying that they were in Joiner's employ to promote the well.

And all this while, no one bothered to see if there was any activity going on at the Daisy Bradford 3. Columbus Joiner had been wise indeed.

Between Laster and Doc Lloyd there had been instant mutual dislike. When Lloyd had arrived to assume command of the well, he had with him a man named Lemmon whom he introduced as a driller.

"From now on, Laster, you can work at night," Lloyd had said crisply. "We'll rig up a generator so you can have some light. Lemmon will take over days."

He had questioned the accuracy of Laster's log-keeping, and they had quarreled about that. He had been skeptical when Laster had explained that 1,800 feet of the hole would have to be reamed—increased in diameter with a larger cutter—before a drill-stem test could be made. The bottom 1,800 feet had been drilled with a bit that had cut a hole six and a half inches in diameter, not large enough to accommodate a testing tool.

They were saved from further quibbling when Laster received word that Daisy Bradford wished to see him in the Just Right Hotel in Henderson. He drove to town. Miss Daisy, who was with her brother, Clarence Miller, was upset. Doc Lloyd had been to see her, she said. Lloyd had been to Van field and had learned a way of completing the well which entailed setting pipe of various diameters in the hole. "He says it's a new technique," Miss Daisy said.

"It's crazy as far as we're concerned," Laster said with some heat. "We'll junk that hole just like the others were junked if we try that. Maybe his way is a good way, but not with the people we've got. Not with the junk we've got."

"He's a smart man," Miss Daisy said.

"We'll junk the hole," Laster said doggedly.

"Mr. Joiner thinks it's best to do it Doctor Lloyd's way."

"We'll junk the hole," Laster said again.

Miss Daisy turned abruptly and left the room. There was some discussion about Joiner while she was gone. His lease on the farm was to expire the next day, Miller pointed out to Laster. Miss Daisy was unhappy because Joiner had not kept up with his lease rentals. She was thinking seriously of not extending the lease agreement as Joiner hoped—and thought—she would do. It still rankled her that she had given up a portion of her royalty to Joiner when they had signed the lease agreement in 1925.

When Miss Daisy returned she walked to Laster and put her hand on his shoulder. Smiling at him, she said, "Now I'll tell you, I'm going to stay with the breeze that carried me this far. If I give Mr. Joiner an extension, I want it set in the agreement that Ed Laster passes on the materials used and supervises the completion of the well."

There was a meeting the following night in Joiner's Dallas office. Present were Joiner, Laster, Clarence Miller, J. H. Shelton, the stockholding automobile dealer, and a lease-hound who ran errands for Joiner, A. W. Brown.

Daisy Bradford had given Joiner a ninety-day extension on his lease, but it was made out in three thirty-day periods. Shelton had been appointed to hold the extensions, and the lease could be terminated at the end of any thirty-day period by action of Shelton or Miss Daisy. This provision was to insure that Laster would be in control of the well until its completion, and the extension document so stated. Joiner was to supply him with the necessary materials. And the old wildcatter was to sign over to Clarence Miller a hundred-acre parcel of the lease.

Joiner read the document. He was now seventy years old. He had spent more than three years trying to drill a well to his "ocean of oil." No one but he knew what sacrifices he had made, what rules he had broken, what chances he had taken. He knew that many oilmen considered him a small-time hustler. He knew that people in the very room he stood in believed that he didn't care one way or the other if he found oil as long as he could sell leases at a profit.

But he knew his own heart. He looked up from the paper and his gray eyes bored into the eyes of Ed Laster. "Ed," he said softly, "you're working against me."

"Not if you want a well," Laster said.

Joiner looked at the clock on the wall. It said two minutes until midnight. The lease expired at midnight.

"You planned it this way," he said, still softly, to those about him.

No one denied it.

He bent over the desk and quickly signed his name to the document. Then he turned and walked out of the room.

Laster had taken the core from the well on July 20. He had made his deal with the Mid-Kansas Oil and Gas Company on or about July 21. Between July 28 and August 7, Mid-Kansas quietly leased up some 1,100 acres in the Daisy Bradford 3 area, paying approximately one dollar per acre.

chapter six

ENTER H.L. HUNT

Laster returned to the drill site to discover that Doc Lloyd and Lemmon, the driller, had sunk the well some seventy-five feet deeper during his absence. He had been angry that the big man had doubted the accuracy of his log-keeping. Now he was furious that Lloyd had tampered with his well. By drilling deeper Lloyd had risked boring into salt water, which would have destroyed the well. That he had not struck salt water Laster counted as another miracle. Nonetheless, he drove to Henderson to confront Lloyd. Lloyd, however, had left the area, apparently for Fort Worth. Laster went back to work, to ream the remaining 1,800 feet preparatory to a drill-stem test.

Meanwhile, Joiner had been in touch with M. M. Miller of Dallas, formerly a lawyer and oilman of El Dorado, Arkansas. Joiner had been told that Miller and his brother Charles had invented a drill-stem testing tool and were anxious to try it out. Ever on the alert for a bargain, Joiner suggested that the Millers not charge him for use of the tool since he would be providing the well. The Millers good-humoredly agreed. They would stand by for a call from Joiner.

There was no shortage of crewmen nor boiler fuel as Laster reamed the hole. Neighboring farmers had heard from the more or less steady crew members that the well was near completion, and ten or twelve showed up daily at the site. The oil fraternity, however, stayed away.

But in Dallas, Bert Ryan, an official of Shell Petroleum Corporation, was having second thoughts about the Joiner venture. He had received reports on activities at the well from his scout in Tyler, Boots Fulton. He had laughed with Fulton about the salted piece of core Joiner had given Reese and McMahan. Now he was wondering if the piece of core had indeed been salted. Wouldn't Joiner have expected oilmen to believe it salted since he gave it away so freely?

Ryan picked up the telephone and called Dale Cheesman in Pecos, Texas. Cheesman was Ryan's utility man; he was well-versed in almost every phase of the oil business. Ryan told him, "Get over to Henderson and take a look at that Joiner well. There's a rumor the driller took a good core. I want to know the truth."

Cheesman arrived at the Joiner site the next day. Laster had put a rope barrier around the well area to keep the farmers from getting too close to the rig. When Cheesman got out of his car and saw Laster on the rig floor, he almost got back in the car and drove off. Eight years earlier, Cheesman, then a drilling contractor in Louisiana, had fired Laster for drinking on the job. He had heard that Laster had married and had quit drinking since that time, but he could not be sure that Laster would not still resent him.

But when Laster saw Cheesman standing by his car, he waved for Cheesman to come and join him. Cheesman ducked under the rope and mounted the rig floor. Laster shook his hand and greeted him cordially. Perhaps Laster was recalling that although Cheesman had fired him he also had promoted Laster from roughneck to driller. Or perhaps he was happy to see a true professional driller whom he respected; he always was lonely for that kind of companionship.

"Have you got any money to buy leases around here?" he asked Cheesman.

"Hell, no!" Cheesman said with a grin.

"Has your wife got any diamonds you can hock?"

"Just one, and she's not about to let me hock it. Why?"

"Because I got a hell of a good core here, Dale," Laster said. He waved a hand impatiently. "Oh, I know. Nobody believes it, but I tell you it's here—and you know I know." He pointed to a bucket. "There's part of it."

Cheesman knelt down and saw an oil-saturated piece of core. He broke off a piece and wrapped it in his handkerchief. "I'm going to tell my people about this," he said to Laster.

"I don't care if you do," Laster said.

Cheesman drove to Henderson and called Ryan. He told him what had occurred. "Will he let McLeod look at the core?" Ryan asked. Angus McLeod was the chief geologist.

"I think so," Cheesman said. "If he doesn't, I have a piece I picked up."

McLeod came to Henderson and Cheesman took him to the well. Laster freely discussed the well with McLeod and let him take a piece of the core. McLeod returned to Dallas, and the next morning Cheesman got word from Ryan to start buying leases at $2.50 per acre—or less.

Shell, like almost all of the large oil-producing companies, had been badly hurt financially by the Depression. The company had adopted a policy of retrenchment. Gulf also lacked funds to buy proved or unproved acreage to any extent. The Texas Company had shifted its emphasis to refining and marketing and was not developing production in Texas as aggressively as it had in years past. Of the Big Four in Texas, only Humble had the means and the will to invest heavily in areas considered promising.

So when Cheesman, crippled by a small purse, began leasing in Rusk County, Humble landmen bird-dogged him wherever he went. They did not immediately begin a leasing campaign, but as a contingency measure they did begin mapping areas they had previously ignored. They had no idea what had prompted Cheesman's unexpected activity, but they wanted to be ready if and when they did find out.

As he drove over the country lanes, Cheesman had time to think of Laster—with gratitude, of course, but also with puzzlement. Why had Laster given him the information and the core? He finally decided that Laster had acted because he wanted a peer to know what he had accomplished against such great odds, that only a peer like Cheesman could appreciate it. Laster had said, "Oh, I know. Nobody believes it, but I tell you it's here—*and you know I know.*"

Laster reamed the well to 3,486 feet. The Millers shipped in the drill-stem testing tool. Only Joiner and a few of his Dallas friends were present when Laster began the test. The tool was attached to the hollow drilling pipe. It was designed to open when it rested against a shoulder at the bottom of the hole. As the tool opened, a vacuum would be created and the mud and water in the well—and the oil and gas if they were present—would rush into the tool and up into the drilling pipe. The tool would close when it was lifted from the well bottom. The fluid in the drilling pipe could be examined as the pipe was pulled from the hole and "broken off" in "stands" of four joints each.

The tool had gone only a short way down into the hole when it hit a tight spot and opened prematurely. Laster pulled it from the hole.

"I'll have to ream the hole again for tight spots," he told Joiner. "I'll let you know when I'm ready again."

He reamed the hole again. But before notifying Joiner in Dallas, he decided to run the tool back into the hole to see if the hole was free of tight spots. He did not prepare the tool for testing. Nevertheless, the tool went to the well bottom and opened briefly—long enough so that when Laster pulled the pipe from the hole it contained some fluid. And the thick, muddy water was cut with oil and gas. Whether or not he had intended to do so, Laster had made a test of sorts. And he now knew that the Woodbine not only was saturated with oil but that the oil most likely could be produced.

The definitive test took place on September 5, 1930. A goodly crowd of farmers was present, along with some oil scouts and geologists. Word about the test had gotten out. Joiner was there with some of his Dallas friends, and so were folks from Henderson and Overton who had interests in the well. Charles Miller, co-inventor of the tool, was on the rig floor to help Laster and his crew.

Standing near Joiner were two strangers who had driven over from El Dorado at the invitation of M. M. Miller, the lawyer-oilman brother of Charles Miller. The strangers were there on a gamble. If the drill-stem test indicated that the well would produce, they intended to offer to finance the casing of the well for a share of its production. Casing was permanent pipe cemented in the hole to serve as the well walls. It would have to be placed before the well could be produced.

One of the men was P.G. (Pete) Lake, an El Dorado clothier and moneylender. The other was H. L. Hunt, who had become an oilman because he had found that oil was the greatest gamble of all. Almost immediately, Joiner struck up a conversation with Hunt, letting him know at one point that he already had arranged for a string of used casing. The conversation was the beginning of a strange but sincere friendship.

On the rig floor Laster lowered the tool to the bottom of the hole. In less than three minutes the smell of gas had raced upward. "That's enough," Miller said—and Laster be-

gan lifting the pipe with its load of mud, water, gas and—
hopefully—oil.

As the joints were pulled out of the hole and "broken
off," the equipment on the derrick floor began vibrating,
shaking the entire creaking structure. As the machinery
rocked against the heavy sills holding it, the noise could
be heard for a quarter of a mile.

Forced upward by the gas, mud and oil suddenly shot
out of the pipe high into the derrick. Then the pipe was
empty, the machinery settled back on the sills, and all
was quiet.

Several people rushed to the rig. "Whaddaya think, Mr.
Laster?" someone shouted. "Whaddaya think?"

Laster shouted back, "It ought to make a pretty good
well—if we can bring it in!"

Some turned to look at Columbus Joiner. He was lean-
ing against a tree, his eyes closed. He opened his eyes
and the smile that was a benison came on his face. The
farmers went to him, holding out their hands in con-
gratulation.

"Not yet," Joiner protested mildly. "It's not an oil well
yet." But he shook the proffered hands.

It was an oil well to the farmers and town folks, and
they wasted no time in sending the news throughout the
county. The word also spread quickly through oil company offices and through the camps where men drilled
for oil. Within a week thousands of eager strangers
swarmed to Henderson, and almost overnight a village of
clapboard shacks and lunch stands sprang up where the
dirt road to the well met the Henderson–Tyler highway.
It was called Joinerville. Cots in the shacks rented for $2
and $3 a night. Hamburgers, which could be bought at
the Depression price of six for a quarter in saner areas,

sold for a quarter apiece. The Miller family's Just Right Hotel in Henderson had four and five men sleeping in each room, with each man paying as if he were the only tenant.

Rumors of fantastic prices being paid for leases spread throughout the county. The leases were being grabbed up by wildcatters and small independent operators. More than 2,000 instruments of sale or lease were filed in the Henderson courthouse within twelve days of the drill-stem test.

But the major oil companies and the larger independent companies still held back. Only Humble made a move: it bought the Herbert Oil Company's interest in the London block, giving Humble and Gulf 4,500 acres in the area of the Joiner well.

Two independent oilmen, Neville Penrose and Edgar Hyde, had been present at the drill-stem test and had been impressed by it. They had hurried to Henderson and called W. W. Zingery in Fort Worth. Zingery was becoming known as the premier land-ownership mapmaker for the oil industry. To lease the land they wanted, Penrose and Hyde needed ownership maps of southwestern Rusk County, and they knew how to get them quickly.

Zingery arrived in East Texas that night. Penrose and Hyde took him to the living quarters of a Humble landman. Humble's landmen had been mapping southwestern Rusk County since Shell's Dale Cheesman, on Laster's information, had begun picking up leases some weeks before the drill-stem test. While Penrose and Hyde entertained the Humble landman with bootleg whiskey in the living room, Zingery sat at the kitchen table and surreptitiously traced the Humble land-ownership map.

Early the next morning, Zingery was at the Henderson courthouse filling in his map. That night he gave the sheriff $20 to let him slip into the closed courthouse. Working around the clock, he made the map as up-to-date as possible. There wasn't time to have the map printed, so Zingery had thousands of photostats made. The first ones went to Penrose and Hyde. The rest he sold—and he sold more almost as rapidly as he could get them made. He worked eight consecutive days and nights without going to bed; one morning he sold more than six hundred dollars' worth of maps while he was trying to eat his breakfast. There were days when he took in more than $5,000.

It was these Zingery land-ownership maps the "little men" used to lease up large sections of southwestern Rusk County while the majors and larger independents ignored the lease play.

While Zingery was making and selling his maps, he acquired four certificates in the third Joiner syndicate. These certificates were different from the ones in the first and second syndicates. Each had a face value of $100 and gave its holder 1/300 interest in the well itself and the eighty-acre tract on which it sat. In addition, each certificate owner was entitled to 4 acres out of 320 set aside by Joiner for this particular syndicate. This meant that only the first eighty certificates issued were valid.

One Saturday morning in the courthouse Zingery met a young Dallas lawyer who was inquiring of the clerk about Joiner's properties. His firm had been retained by the Cleveland Trust Company because some Joiner certificates had been found in the estate of a customer who had died. "Looks to me like this man Joiner sold more interests than there is interest," said the young lawyer. "This might call for a federal receivership."

Like most oilmen and those associated with the industry, Zingery abhorred federal intervention in local matters. He went to John Alford, the Henderson Ford dealer, who he knew held some of the certificates. He told Alford about the lawyer's remarks, and added, "Federal receivership would be a bad thing. It would take control of this whole deal away from local people. Why shouldn't we get a local receivership?"

Local receivership was a good idea, Alford said, and he

telephoned his attorney, Tom Pollard of Tyler. Pollard drove to Henderson right away. Because it was a Saturday afternoon, the courthouse was closed. But Pollard telephoned Judge R. T. Brown at home, and Judge Brown met them in his chambers. A receivership action, Zingery versus Joiner, was filed, and Judge Brown set a hearing for thirty days in the future.

It was this kind of action that Joiner had anticipated when he had signed the lease-extension agreement two minutes before midnight in his Dallas office, and when he had leaned against a tree with his eyes closed at the end of the successful drill-stem test. For Joiner had greatly oversold the three syndicates, and he had sold some leases several times, one to eleven different buyers. By bringing in the well—if it could be completed—he would be fulfilling his promise to the farmers while endangering his own freedom and practically insuring the ruination of his reputation.

"You're working against me, Ed," Joiner had told Laster at the midnight Dallas meeting.

"Not if you want a well," Laster had replied.

And Joiner had signed the papers. He had decided to take what punishment might come.

But the Zingery suit had hardly been filed when the rumor began spreading that it had been instigated by the major oil companies. The rumor had it that the majors aimed to "gobble up the field" during the confusion created by the lawsuit and others which were certain to follow. Doc Lloyd, who had returned from Fort Worth, called the rumor to the attention of several newspapers. The response was astonishing.

Stories praising Joiner's courage and perseverance appeared on front pages around the area. He was called the Daddy of the Rusk County Oil Field, and was referred to as "Dad" Joiner, a name he bore thereafter. Carl Estes, the fiery editor of the Tyler *Courier-Times*, wrote an editorial denouncing the "slick lawyers" Estes saw moving in on the old wildcatter.

Estes wrote, in part:

> Is he [Joiner] to be the second Moses to be led to the Promised Land, permitted to gaze upon its "milk and honey," and then denied the privilege of entering by a crowd of slick lawyers who sat back in palatial offices cooling their heels and waiting while old "Dad" worked in the slime, muck and mire of slush-pits and sweated blood over his antiquated rig, down in the pines near Henderson? Is it right to tie this poor old fellow up in court while the horde of "big boys," which has flocked into the territory adjacent to his holdings, sinks holes and sucks the oil which he discovered, right out from under him?
>
> We are sure that we voice the sentiment of the masses when we say emphatically—No!
>
> If they are permitted to tangle him up in court, we predict that other men of small means, land owners and farmers, will follow him to the guillotine of financial execution. East Texas has had enough of inquisitious "oil trusts" without sitting quietly by any longer while the "big boys" crowd in, leaving "Dad" Joiner out of the picture.
>
> Come on, congressmen; come on, governors; come on, railroad commissioners, state senators and representatives. Every one of you to a man was bawling around from every stump in this state, preaching the doctrine of "retrieving the Little Man, the home and small farmers and busting the trusts." This is your opportunity to make good on those sweet-sounding promises.
>
> It is high time the independent oil operators had their inning.
>
> And lastly, if they haul "Dad" Joiner into court, you

farmers and small land owners should turn him back his
oil well.

Now, if this be bolshevism, then we're bolshe-
vists . . .

Such editorializing could not help but raise support for
Joiner, but by calling attention to the lawsuit it awakened
suspicion among groups of investors who had been waiting
passively for their dividends to begin rolling in. Thus there
was talk of other suits in the offing; more rumors of intrigue
and black deeds were circulated.

The rumors and the editorials did not try to identify the
"big boys" who had flocked in to sink holes and suck the oil
right out from under Joiner—and it would have been an
impossible task. Only Shell was leasing in southern Rusk
County, and its landmen were so limited by a tight budget
that they could not compete with the freewheeling independent operators. Despite the drill-stem test, most major
oil company scientists and officials still did not believe that
oil would ever be produced in commercial quantities in
Rusk County.

One solid piece of news emerged during this period of
confusion: a "Joiner Jubilee" would be held in Overton on
September 22, and all were welcome to come and pay their
respects to the Daddy of the Rusk County Oil Field.

So in this first year of the Great Depression, there were an
oil boom, a brewing court fight and a great oil celebration
planned in an area where not a single well was producing
a single barrel of oil.

At the Daisy Bradford 3 work was going smoothly, although it didn't appear so to the impatient farmers. Joiner
had arranged for three storage tanks and a string of used
casing from the Standard Pipe and Supply Company. The

wildcatting son of an old friend in Ardmore had loaned him some excellent machinery to mount under the rickety old derrick. D. H. Byrd, called Dry Hole because of his initials and because of his luck—he had drilled fifty-six dry holes in succession before discovering two oil fields on the same day —had the equipment shipped in from Elkhart, Texas. The young wildcatter liked and admired Joiner and didn't want to see him lose the well for want of adequate machinery.

Doc Lloyd and Laster had made an uneasy peace. Graciously, Lloyd had apologized for his earlier conduct and Laster had accepted. Laster had lots of help now. W. A. Kirkland, a driller, was working with him. Besides Glenn Pool and Dennis May, he had as helpers Dave Cherry, Jim Lambert (the burned fireman), Dave Hughes, Pete and Jake Maxwell, Fred Newman, J. W. Duncan and J. Sistrunk. Most of them had worked intermittently on all three wells.

H. L. Hunt, the El Dorado oilman, was a frequent visitor. He had not returned home but had found living space in Henderson. He had been introduced to Joiner as Haroldson Lafayette Hunt, but the old wildcatter called him "Boy." The two spent hours talking, enjoying each other's company.

Hunt was forty. Like Joiner he did not drink or use profanity. Like Joiner, he was unusually attractive to women. The waitresses in the jerry-built cafes in Joinerville and Henderson vied to serve his meals. The public stenographers who had set up shop in clapboard offices were quick to take his dictation and type his letters. Women long married turned their heads to watch him pass on the board sidewalks.

Hunt was tall and broad, running a little to fat at his girth, and was what the waitresses called a spiffy dresser. His trousers were always pressed, his shirts always starched, his necktie always properly adjusted, his sailor strawhat always at a jaunty angle on his blond head. He smoked big cigars (which Joiner abhorred). Though he appeared to be well-off financially, he was in fact short of cash, a condition that had never worried him in the past and did not disturb him now.

He had raised enough money, however, to lease four tracts as near the Joiner site as possible. He had acted on the strength of the drill-stem test. One tract lay to the south and the others to the east of the Joiner test.

Hunt was born on a farm near Vandalia, Illinois, in 1890. His father was a farmer and grain speculator. He did not attend school a day in his life, and left home at the age of twelve. He hoboed throughout the West, working occasionally as a farmhand or a cowboy, then moved north into Canada, where he worked as a lumberjack. Such a life at such a tender age sharpened the youngster's already keen mind, and it was during these early years that he learned to "count the cards," a requisite in poker and other card games.

He was not yet twenty when he wandered into Arkansas, drawn there by stories his father had told him of the land's richness. The elder Hunt had fought in the battle of Ditch Bayou, Arkansas, during the Civil War. With the considerable money he had accumulated through hard work, ingenuity and good luck, young Hunt became a cotton planter near the community of Lake Village. By 1912, when he was twenty-two, he was, as he told Joiner, "owner of lots of land and lots of liabilities." The price of cotton had dropped dramatically and was slow in rising, so Hunt veered into the buying and selling of plantations in the Mississippi Delta country.

He had made a fortune by 1919, but he watched it disappear in the depression year of 1920. It was at this disastrous turn in his life that Hunt heard about an extraordinary event: A great oil gusher, the Bussey well, had blown in wild at El Dorado and was spewing thousands of barrels of oil daily into the sky.

Hunt was in El Dorado before the well was brought under control. Broke, he had borrowed $50 for transportation and "walking-around money." He entered the oil business much like Joiner had done. He would drive out in the country, offer a farmer $25 per acre for a lease, then drive back to town and find someone willing to pay $35 per acre. Then he would pay the farmer and keep the $10-an-acre profit.

By such dealings and by his occasional skill with cards, Hunt was ready to drill his first well in July 1921, six months after arriving in El Dorado. The well, in the El Dorado South field, was a success, and so were others that followed. In the meantime he sometimes gambled in a casino and in the Garrett Hotel, where a no-limit poker game was continuously going on. By 1925 he was able to sell a half interest in some of his properties for $600,000.

But as he told Joiner during one of their long talks, Hunt liked to "move his money around." By September 1930, when he came to Rusk County, Arkansas appeared to be pretty well drilled up. Hunt owned a family home valued at $125,000 in El Dorado, but the Depression had gnawed deeply into his assets. He was ready for something new.

Joiner was not surprised that a man with Hunt's background had learned "to turn a pretty phrase," as Joiner put it. And he was not surprised that Hunt was as well-read as himself. Each recognized the other as a supersalesman.

Joiner could not hide his desperate financial situation from a man as perceptive as Hunt, and Hunt could not avoid hearing rumors of the wildcatter's dubious transactions. The two men did not discuss this matter in detail—but Hunt had seen Joiner's reaction to the drill-stem test. To Hunt, this reaction verified the rumors.

The "Joiner Jubilee" was the brainchild of R. A. Motley, the Overton banker. Motley was genuinely grateful for the prosperity the show of oil in the Joiner well had brought. He also was aware that Henderson was getting the lion's share of the increased business. The celebration would demonstrate the people's gratitude and at the same time focus attention on Overton as the logical headquarters for the oil activity. He explained his views to Walter Tucker; Tucker went to work with his usual enthusiasm for a new project.

On the afternoon of September 22, Overton's main street looked like a carnival midway. Streamers bearing Joiner's name hung across the street. Boys and girls raced up and down the board sidewalks with pennants. Beautiful floats depicting oil-field scenes moved slowly along the unpaved street. Games were organized on vacant lots. Oilmen gathered in knots to talk business. And from great pits at the edge of town floated the aroma of barbecuing beef, pork, chicken and mutton.

At eight o'clock that evening several thousand persons passed by the long tables laden with food and moved out to get comfortable on the grass. Fiddlers played, glee clubs sang. Doc Lloyd walked amidst the throng a hero. His hand was shaken, his back was slapped.

Joiner was introduced. Cries of "Hooray for Dad Joiner" accompanied the old wildcatter to the platform. He looked very tired, but his smile was as bright as noon-time snow. He said only a few words: "I thank all of you for coming here. Nothing could have been done without your help . . ."

They honored him, showered their love on him, reached out to touch him as he moved among them—and his magnificent eyes were shiny with tears.

chapter seven

THE DAISY BRADFORD 3

They came on foot and on horseback, by buggy, wagon, truck and automobile. They had heard the news by word of mouth. Neighbor had told neighbor; the rural telephone lines had jangled with excitement. Rural mail carriers had passed the word along their routes. Oilmen had received the information from scouts, landmen and lease-brokers. By nine o'clock in the morning of October 3, 1930, more than 8,000 persons had tromped their way into the Daisy Bradford 3 clearing, and hundreds more were churning up the dusty road from the paved highway. It seemed almost as if Gabriel had blown his call to judgment in this faraway woodland and 10,000 had responded.

Miss Daisy had wanted the well brought in on October 2, because that was Caroline Laster's fourth birthday. But the cementing of the casing had not been completed until the afternoon of that day, and it was the next morning before Laster was ready to drill out the cement plug at the bottom of the casing so that bailing could commence. The delay had given even the most laggardly an opportunity to be present.

It was like a religious revival or a great camp meeting. The throng was optimistic, enthusiastic and good-natured. Bootleggers mixed with the devout and sold pints of white lightning and pop-skull, which helped keep the chilling wind from biting too deeply into some. And Jim Hale Miller, Miss Daisy's enterprising teenaged nephew, sold dozens of cases of iced soda pop and scores of homemade sandwiches from behind a small stand he had erected in a grove of trees.

"But where's Dad Joiner?" someone asked, and the question went the rounds. Joiner was not to be found. Ed Laster could not be asked; he and the crew were busy sending the bailer into the well and bringing it up laden with mud and water. By relieving the pressure of the mud and water on the oil sand, Laster hoped to coax oil to the surface. But the oil did not come.

Word spread that Joiner was ill in Dallas, and a moan of sympathy sounding like wind through the trees burst from the crowd. But Joiner was not really ill. He was aware of the ever-growing suspicions about his dealings, and he didn't want to face those who might look on him with loathing and contempt. At the very time the well might be coming in, he didn't want to answer questions as to how it had been accomplished.

All through that day Laster and his crew ran the bailer without a show of oil. When darkness fell, lanterns were lighted and the work continued until the crew was exhausted and Laster signaled for a halt.

When the work resumed the next morning, the people were there. If the crowd had thinned, it could not be detected. Young Miller had rounded up more soda pop and sandwiches, but he also had gained some competition. Two grown men were hawking food and drink from the back of a pickup truck. There were customers enough for all three merchants.

By ten o'clock Laster had bailed the hole as free of mud and water as possible, and still there was no show of oil. At that moment, as if on signal, a murmur went through the crowd. "Dad Joiner's coming through," someone said, and they made way for their hero. They patted him on the back and shouted welcomes to him as he wormed his way through the vast assembly of people and vehicles. He returned the cheers with a smile and a wave.

He climbed through the wire fence Laster had now erected to keep the people from getting too near the derrick. He mounted the rig floor, and the crowd cheered him as he conferred with Laster. They continued cheering him as he left the floor and joined Dry Hole Byrd, his young wildcatter friend who had supplied the new machinery for the rig.

Laster and the crew began "swabbing" the well. The swab was a steel and rubber device which was attached to a steel cable and lowered into the hole. It fitted snuggly inside the casing, and was open at the bottom so that it could go through mud and water. After it had been lowered to the depth Laster determined proper, he would begin to draw it from the hole. Its bottom opening would close and the vacuum created was supposed to pull the oil from the Woodbine. But with each swab the word that came from the rig floor was "Nothing yet."

Night came and the lanterns were lighted again. "Nothing yet," Laster said as the swab again brought mud and water flying from the hole. Again he shut down the rig for the night.

The next day, October 5, was a Sunday, and daylight found many of the devout on their way to churches to pray for the well's success. Still, there were never less than 5,000 present during the day as Laster and the crew continued swabbing. Joiner was glassy-eyed with fatigue and strain. Laster and the crewmen were short-tempered. The professional oilmen had long ago decided that if the well did come in it would not be a big one. But the farmers and the town folk never wavered in their faith. When firewood for the boilers was exhausted, farmers ripped tires from their old trucks and offered them as fuel. The stench of burning rubber rose to high heaven, but the work went on.

It was late in the evening when Laster, drawing the swab from the casing, heard a gurgling sound deep in the hole. He spun around, shouting, "Put out the fires! Put out your cigarettes! *Quick!*"

Those nearest the rig felt a slight trembling of the earth. The gurgle became a roar. Suddenly a column of oil and water shot high above the derrick. It spread out like a titan's umbrella and fell down upon the pressing crowd like a torrent of raindrops.

And the crowd went mad. Men held their faces to the sky, shouting into the oil that sprayed them. They rolled in the oily dirt of the clearing. They hugged each other and each other's wives. Dennis May, the colorful crewman, jerked a pistol from his pocket and began firing into the oil umbrella. "Little Red" Selby, a Shell scout, slugged May behind the ear and two men standing nearby grabbed the pistol from his hand. A single spark could have exploded the well by igniting the highly volatile gas.

The roar abated but the oil continued to spout. When he decided it was free of mud and water, Laster spun valves and directed the oil into one of the tanks. There was quiet. Joiner asked Dry Hole Byrd to gauge the flow. "Whisper it to me," he said.

Byrd got his gauges and went to the tank. When he returned he whispered in Joiner's ear, "She's flowing at the rate of sixty-eight hundred barrels a day!"

Joiner gasped. Without thinking he shouted high above the noises of the crowd, "SIXTY-EIGHT HUNDRED BARRELS! UNBELIEVABLE!"

Leota Tucker, who had cooked and scrubbed and scrounged and lived in a tent during the drilling of the first two wells, grasped the arm of a man next to her and said, "I'll never wring out another dishrag in my life!" The man was J. Malcom Crim, who had been trying to get someone to drill his farm near Kilgore since the fortuneteller had prompted his dream of oil a decade earlier. He patted Mrs. Tucker's hand, but his eyes were wistful.

Joiner's shout had let the cat out of the bag. This was no dinky well, it was a whopper. Everyone wanted to spread the word. The crowd dissipated as if it had been promised the sight of another gusher a few miles distant. This was it: the impossible had happened. Rusk County had its first oil well.

Dad Joiner had kept the faith.

The next day it began raining as if the gods intended to make up for the years of drought. It was a "frog-strangler" of a rain that turned the roads and fields into quagmires. A farmer stomped his booted foot in the lobby of the Just Right Hotel. "I'll be damned!" he snorted. "If it had rained like this the past five years I wouldn't give a hoot if I'd never seen Dad Joiner!"

chapter eight

THE BIG DEAL

Five days after the well came in, most officials and scientists of the great oil companies had marked off the Daisy Bradford 3 as a freak. Its settled production was about 250 barrels per day, not 6,800—and it was flowing by heads. It would flow some 180 barrels, then cease flowing for eighteen or twenty hours before flowing again. It seemed as if every gasp would be its last.

The promoters and nonprofessionals and independent oil-men, however, were not overly concerned about the well's behavior. They swept across southern Rusk County and into neighboring Smith County, buying and selling leases. Every day brought reports of wells soon to be drilled. Every day brought freight trains loaded with roughnecks and roust-abouts from other oil provinces into Henderson and Over-ton. Men were sleeping in vegetable loading sheds at rail-road sidings from Troup to Tyler.

An Oil Exchange was opened in Henderson to expedite

lease trading. Abstractors, lawyers, and stenographers were kept busy through the nights preparing leases and sales contracts or writing reports. A typical person on the muddy streets of Henderson or Overton had a checkbook in one hand, a map in the other.

But the Daisy Bradford 3 itself fascinated H. L. Hunt. He had been present when the well came in, and he had stayed on to witness its peculiar behavior. From the time of the drill-stem test on September 5 he had been studying the general area and its oil history. Now it came to him that the well flowed as it did because at the bottom of this well the Woodbine had thinned out so that each head of oil had to come from a thicker parent sand body, possibly through stringers of sand, before accumulating in the thin sand at the borehole. This would mean that the Joiner pool—if he could call it that—was fed by a much larger pool of oil.

Hunt had learned of the dry holes sunk by earlier-day drillers to the east of the Joiner well. It followed, then, that the larger pool of oil he thought was present would lay to the west.

It will be recalled that Hunt had acquired four leases soon after the drill-stem test—three to the east and one to the south of the Joiner well. Now he decided to ignore the eastern leases and drill on the south lease. But what he really wanted was to acquire leases to the west and north of the Daisy Bradford—and Joiner held shaky control over what appeared to Hunt to be the most promising.

Hunt also had another iron he wanted to put into this intriguing fire. Joiner was already contracting to supply oil to others planning to drill nearby leases. The oil would be used as boiler fuel. The oil had been tested for quality; it was 37.3 gravity, which meant that it was almost free of sulphur and other undesirable elements, and that it was a light oil readily refined to produce gasoline and other products remunerative to the refiner.

Hunt immediately decided that if he found oil it would

not be wasted as boiler fuel. He laid plans to build a four-inch pipeline from the vicinity of the Joiner well to a point three miles away on the Overton–Henderson branch line of the Missouri-Pacific Railroad. He had built and operated "gathering" lines in Arkansas, so he knew the ropes. He met with Sinclair representatives; the company had a refinery in Houston. Eager to purchase high-quality oil, Sinclair agreed to build a loading rack at the rail point. Hunt knew he could find plenty of pipeliners in Henderson to begin work on the project.

Then he turned to the drilling of a well on his south lease and whatever plans he was formulating for acquiring leases on to the west.

Among the boomers who had romped into Rusk County after the Joiner drill-stem test were P. S. Groginski and Ed Zilkey, employees of a poor-boy oil company headed by Ed Bateman. Groginski was a geologist. Zilkey, a former minor-league baseball player, was an oil jack-of-all-trades and competent at all. Bateman, their boss in Fort Worth, was a former newspaper-advertising salesman who had worked on the Houston *Post* and Dallas *Times Herald.* No one outside of Dad Joiner ever wrote more glowing copy about a well to be drilled in wildcat territory.

Groginski went out from Henderson to inspect the Joiner lease. Zilkey hung around Henderson to hear what he could hear. He ran into an old driller, Elmer (Mutt) Hayes, who was living on a farm north of the Joiner site. Zilkey and Hayes had mutual friends; both had worked in the Arkansas oil fields.

Hayes told Zilkey there were three places to lease and drill in all of East Texas: a spot in Smith County, another near Gladewater in Gregg County, and "some land up near Kilgore owned by the Crim family."

"Do you promise to give me a job as driller if you take one of 'em?" Hayes asked.

"I promise," said Zilkey.

Groginski and Zilkey checked out the three localities, arriving last in the Kilgore area. By then they had learned of the dry holes to the east of Joiner's well and the salt water wells in Smith County to the west. They wanted land either north or south of Joiner's site, and the Lou Della Crim farm fit the bill.

They went to the Crim store in Kilgore to see Malcom. They were told that he was in Henderson trying to get involved in the lease action. "You'll recognize him," Mrs. Crim said. "He'll be smoking a corncob pipe."

Groginski and Zilkey found Crim on Henderson's main street. Crim did not immediately display his eagerness to get the Crim farm drilled. He told Groginski and Zilkey he was offering leases on 4,000 acres at $2 per acre. He may as well have been asking $200 per acre as far as the two wildcatters were concerned; they were living on a shoestring.

Zilkey shook his head impatiently. "We're not here to talk about buying leases. We're here to tell you we'll drill a well on your farm if you can get about fifteen hundred or two thousand acres together. We'll move in a drilling rig soon as the papers are signed."

Crim puffed his corncob pipe and looked thoughtful. He was under no illusions about the Bateman Oil Company. In the past ten years he had learned a lot about all kinds of oil companies. He knew Ed Bateman ran a poor-boy operation. But he also knew that even at this point—with all the frenzied trading going on in southern Rusk County—not a single major oil company had shown the slightest interest in dealing with him. There really was very little to look thoughtful about, so Crim said he would deal.

Back in Kilgore he quickly rounded up the neighboring acreage belonging to the Laird family on one side, the Peterson family on another. All told, the block amounted to 1,494 acres. Groginski and Zilkey returned to Fort Worth to inform Ed Bateman of the details. All three drove back to Kilgore on October 14. The agreement was signed by Bateman, Crim, Ben, Shack and Roy Laird, and John and Ben Peterson. Ben and Shack Laird owned a cotton gin in Kilgore, and the deal was consummated at a worktable at the gin scales. It called for Bateman to drill to 4,000 feet or pay sand, whichever came first.

Three days later Bateman's drilling rig and wooden derrick were moved into Kilgore. Crim's worst fears were realized. The Coffee Pot rig was as dinky as Dad Joiner's and in almost as bad a condition. Like Joiner's, it had been manufactured to drill only to 2,500 feet. But Bateman, like Joiner, was a most persuasive man. "Don't fret," he reassured his new partners. "This thing could dig to China if it had to."

At thirty-five, Bateman was a fine figure of a man. He had Joiner's flair for words, but he was effusive where Joiner was quiet. He awakened sleepy Kilgore with his mere presence. He sent Groginski out to "geologize" the Crim farm while he busied himself at the typewriter grinding out lurid "mailers" to potential investors.

Groginski conducted a quick survey of the Crim property, and drove a location stake at a point he declared was the "center of the oil field." Fortunately, the site was next to a road and near a stream. This meant Bateman would not have to build a road to move in his rig and materials, nor would he have to spend money sinking a water well to supply water for drilling. The site was thirteen miles almost due north of the Daisy Bradford 3.

Zilkey began trying to round up needed materials, trading off small pieces of interest in the well, with Bateman's blessing, to supply houses for bits, pipe and cables. And Zilkey did not forget his promise to Mutt Hayes. He hired the driller as soon as the rig was moved to the drill site.

Bateman also hired as head driller one Bill (Checkbook) Cain, a colorful oilman who had gained his nickname by firing every roughneck on a job in the Wortham field in a fit of anger. Cain had walked around writing out final checks for everybody who worked under his control. Now he and Bateman undertook to wrestle daily at the drill site to see who would make the coffee.

Crim's faith received another blow when a group of Wichita Falls oilmen—lured to Kilgore by Bateman's mailers—took one look at the derrick and drilling equipment and promptly returned home.

"We don't need those guys," Bateman said heartily. "We'll have money coming in here by the carload before long."

The money trickled in, but enough was received so that on the twenty-eighth day of October, twenty-three days after the Joiner well spouted oil, the Lou Della Crim 1 was spudded in. By that time postal inspectors were showing a mild interest in Bateman's prose.

And what of Barney Skipper, the Longview oil prophet? His business was being so ably conducted by the resolute, mustachioed Walter Lechner that Skipper dared hope that his dream might finally come true. Back in September, at the time of Joiner's drill-stem test, Lechner had told Skipper, "I'm tired of trying to sell something to people who haven't got enough sense to buy. I'm going to Fort Worth and see Johnny Farrell."

John E. Farrell was an old friend and business associate with whom Lechner had dealt over the years while Farrell was an official of the Marland Oil Company. Now Farrell was an independent operator with the reputation of a "closer," an extremely meticulous and astute trader. Talks between Lechner and Farrell stretched over a period of weeks. Because of Lechner's persistence, and because Farrell respected and trusted his old friend, he decided to take the plunge. He said he would have a partner in the deal, W. A. Moncrief, also of Fort Worth. Moncrief in turn said he had a drilling contractor, Eddie Showers, who would be interested in drilling the block.

The deal was an interesting one. Lechner and Skipper had leases on 9,300 acres in Lechner's name. Lechner agreed to sell Farrell 5,000 acres at fifty cents per acre with Farrell guaranteeing to drill a well on his acreage but on a site of Lechner's choosing. Lechner calculated that he and Skipper could make a killing selling leases on their remaining 4,300 acres or by drilling the land themselves—if Farrell struck oil.

There was one hurdle: Titles to some of the leases in the 9,300 acres were in bad shape. Lechner, however, was much too methodical not to reduce errors to the irreducible minimum. But he was almost worn out from his work assembling the block and trying to interest oil companies in the acreage. He told Skipper, "I know a man named Ray Hubbard who's an expert at 'curing' titles. Let's call him in. Let's give a piece of our piece to him to clean this thing up."

So Hubbard was called in and put to work "curing" titles. Lechner and Skipper gave him a third of their 4,300 acres to do this special job, but the interest was undivided. That is, all three would profit equally from whatever Lechner could produce from the 4,300 acres.

Meanwhile, Farrell was having his troubles. The $2,500 he had given Lechner and Skipper for the 5,000 acres had almost stripped him. He needed a derrick and drilling rig and all the other necessary equipment. He began dealing with the Seminole Tool and Supply Company for a derrick and drilling rig, and with the Arkansas Fuel Oil Company for drill pipe and other materials.

The entry of these two established firms into the project was heartening to Skipper. Though he trusted Lechner implicitly, he was a bit dubious of Farrell and Moncrief. For one thing, he thought they were breezy fellows; had they not flown to Longview from Fort Worth for their first look at the land? Skipper had warned them by telephone that Longview had no airport, but they had flown in just the same; Skipper himself had waved them in for a perfect landing on the Bevins' pasture, not far from town. Lechner's assurances that Farrell and Moncrief were solid citizens had relieved most of Skipper's fears. Now with two reputable firms involved in the project he was satisfied.

Almost every adult who was in Henderson on October 18 tried to get inside Judge R. T. Brown's courtroom for the Joiner receivership hearing. Joiner was not present; he was reported ill by his attorney, Colonel Bob Jones of Henderson. Dozens of other certificate holders had joined Zingery in the action by now, but their attorneys were allowed only to state the basic thrust of their allegations. Judge Brown would hear nothing more. As soon as the brief statements were made, Judge Brown rapped his gavel and said quietly, "I believe that when it takes a man three and a half years to find a baby he ought to be able to rock it for a while. This hearing is postponed indefinitely."

The postponement set off a brief but lively celebration in Henderson, but Joiner had very little time in which to rock his baby. Zingery, who had been maligned by gossips and newspaper editorialists as a tool of the major oil companies, moved to have his suit dismissed the following day. It was dismissed, but a few days later another receivership suit was filed, this time in Dallas. Advertisements ran in most major Texas newspapers calling for those with claims on Joiner to join in the court action. A hearing was set in Dallas for October 31.

But Joiner was not to be found; the receivership papers could not be served on him. Then Alvin Owsley, an attorney for the plaintiffs, received a tip that Joiner was hiding out in the Adolphus Hotel. The hotel management denied this.

But Owsley handed a crisp hundred-dollar bill to a bellboy. The bellboy gave the lawyer Joiner's room number and the papers were served.

Joiner was pale and drawn when he took his seat at the defense counsel's table with Colonel Bob Jones and a legal associate. As soon as the plaintiffs' attorney announced ready for trial, Colonel Jones rose. "If it please the court," he said, "we wish at this time to present a petition for voluntary receivership on behalf of our client, C. M. Joiner, for the Joiner properties in Rusk County, Texas—"

He was interrupted by a plaintiffs' attorney. They haggled. The judge gaveled. Jones continued, "Mr. Joiner has been so harassed that he feels it now impossible for him either to develop or sell his holdings to the satisfaction and protection of all investors. Hence, he prays that a receiver be—"

Again Jones was interrupted, and again they haggled. But it ended with E. R. Tennant, a Dallas banker, being appointed receiver for the well, the eighty-acre tract on which it sat, and the various leases Joiner had carved out of his block to sell through his three syndicates.

Standing in the wings were a growing number of unhappy citizens who were anxious about their lease deals with the old wildcatter. At this point he was approached by H. L. Hunt who suggested that Joiner sell out to him "lock, stock and barrel."

"Boy," said the old wildcatter wearily, "you'd be buying a pig in a poke."

But he listened to Hunt's offer, and rejected it.

Hunt had erected a steel derrick on his south lease on October 20 and had spudded in the test a week later. Several other operators were drilling also, but Hunt paid them little heed because their leases lay in directions he considered unproductive. But he was greatly interested in a test being sunk about a mile to the west of the Daisy Bradford 3 on the Ashby farm. The well was being drilled by Foster and Jeffries, contractors, for the Deep Rock Oil Company, an independent. The test was on a small lease almost surrounded by Joiner leases.

Hunt knew this test would either prove Joiner's acreage as productive or condemn it. He believed it would be a good well. He held talks with Frank Foster of Foster and Jeffries, and Foster introduced him to a Deep Rock official. Hunt suggested that he and Deep Rock buy out Joiner, but Deep Rock didn't want to deal. He went to several major oil companies with whom he had done business in Arkansas to see if he could interest them in the deal. Still bound by the opinions of their scientists, the companies said No.

"Buy him out yourself," Hunt's clothier friend from El Dorado, Pete Lake, kept insisting.

"If you're so hot, will you take an interest?"

"Hell, yes!"

"How much?"

"I'll take a fourth," Lake said.

Hunt had moved key members of his organization to Henderson from Arkansas. Now he put three men to watching the Deep Rock well—Charles W. Hardin, Jick Justiss and Robert V. Johnson—while he and Lake went to Dallas to try to deal with Joiner. They rented a number of rooms in the Baker Hotel and paid court on the old wildcatter.

Meanwhile, the news from Rusk County was all bad. Arkansas Fuel Oil Company had drilled a dry hole a half-mile southeast of the Daisy Bradford 3. So had L. L. Smith on a lease less than a mile to the northeast. Immediately the bottom dropped out of the lease market. This did not disturb Hunt; he had anticipated the dry holes—and a sagging lease market should make Joiner more amenable to dealing. He kept pressure on the old wildcatter.

Hardin was keeping Hunt informed on the progress of the

Deep Rock well. Joiner spent a great deal of his time in Hunt's hotel room on November 25 and November 26. At 4:30 P.M. on November 26, Hardin called Hunt from Henderson and said the bit had reached the Woodbine in the Deep Rock well.

It was a cold, rainy day in Rusk County. The Hunt scout on the well was Robert Johnson. Just about dark, a core was taken from the well. Johnson, an experienced driller, got a six-inch piece of it. He rubbed it between his fingers and tasted it. He rushed to a telephone and called Hardin in Henderson. Hardin immediately called Hunt in Dallas. The core barrel had cut sixteen feet of material, Hardin said, and ten and a half feet of it was oil-saturated Woodbine. It was about 8:30 P.M.

Some four hours later Joiner signed contract papers prepared by Hunt and his attorney, J. B. McEntire.

In the contract Hunt agreed to pay Joiner $979,000 for Joiner's leases on about 5,000 Rusk County acres, most of it from sales of oil produced. In a separate deal, Hunt agreed to pay Joiner $203,000 for Joiner's interest in the 80-acre Daisy Bradford tract on which the discovery well was located, and $153,000 for his interest in a 500-acre lease elsewhere on the Bradford farm.

The total sales price was $1,335,000.

Hunt made a $24,000 cash down payment on the 5,000 acres, a $3,000 cash down payment on the 80-acre tract containing the well, and a $3,000 down payment on the 500-acre lease.

It was "understood and agreed" that the lands in the 5,000-acre deal "have not been actually surveyed on the grounds, and that the respective acreage listed is estimated to be substantially correct." The section of the agreement detailing the financial aspects of the deal, said:

> The consideration paid and to be paid by the party of the second part [Hunt] to the party of the first part

[Joiner] is Twenty Four Thousand and No/100 Dollars ($24,000.00) cash in hand paid, the receipt of which is by the execution of this instrument and by the party of the first part acknowledged and confessed.

And the further consideration that the party of the second part make, execute and deliver to the party of the first part, his four (4) certain promissory notes bearing even date herewith, bearing interest at the rate of six percent (6%) per annum from date until paid, payable to the order of the party of the first part at the First State Bank, of Overton, Texas, in amounts and maturing as follows:

> One (1) note for $10,000.00 due 90 days after date;
> One (1) note for $15,000.00 due 120 days after date;
> One (1) note for $10,000.00 due 180 days after date;
> One (1) note for $10,000.00 due 270 days after date.

And the further consideration that the party of the second part shall pay to the party of the first part, out of the proceeds of the sale of an undivided seven thirty-seconds (7/32nds) of the oil produced, saved and marketed on or from the lands and premises covered by said oil and gas leases as hereinabove described, the sum of Three Hundred and Seventy Five Thousand, Five Hundred and Fifty Five and 55/100 dollars ($375,555.55), said payments to be made by the party of the second part, or his assigns, to the party of the first part, or his assigns, if, when and as oil from said lands and premises is produced, saved and marketed, and after said Three Hundred and Seventy Five Thousand, Five Hundred and Fifty Five and 55/100 Dollars ($375,555.55) has been so paid, the party of the second part, or his assigns, agrees to pay to the party of the first part, or his assigns, the further sum of Five Hundred and Thirty Four Thousand, Four Hundred and Forty Four and 45/100 Dollars ($534,444.45) out of the proceeds of the sale of seven sixty-fourths (7/64ths) of the oil next produced, saved and marketed on or from the lands and premises covered by said oil and gas leases as hereinabove described, said payments to be made by the party of the second part, or his assigns, to the party of the first part, or his assigns, if, when and as oil from said lands and premises is produced, saved and marketed.

The agreement reiterated that Hunt would make payments only from oil sales from the 5,000 acres, and made clear that Hunt would "be free to exercise his discretion as to the time and manner of development." Thinking of the flood of lawsuits he anticipated, Hunt wrote: "Should title to any lease or leases be attacked in judicial procedures, the party of the second part may at his discretion and at his own expense, defend such title so attacked, but shall not be liable to the party of the first part on account of his failure to so defend or successfully defend such title."

And he wrapped up the matter of dubious titles in these paragraphs:

Should the leases or either of them herein conveyed be now, or later become involved in litigation or other controversy, and as a result thereof payments for oil and/or gas produced, saved or marketed therefrom be withheld, there shall be no obligation on the part of the party of the second part to pay to the party of the first part the proceeds of the sale of such oil and/or gas due or that may become due the party of the first part by the party of the second part as herein provided, unless and until said payments of such oil and/or gas shall have been made by the purchaser thereof to the party of the first part in accordance with his proportionate interest therein as herein stated, to be by him applied on said deferred payments to be made out of the proceeds of the sale of oil.

And for the consideration above herein recited, the

party of the first part . . . binds himself that the represen-
tations and declarations by him made herein are true
and are limited and qualified only by the stipulations in
this instrument contained, and any loss sustained by the
party of the second part, or his assigns, on account of
breach of any representation of the party of the first
part herein contained, shall be a charge and proper
offset against the deferred payments to be made out of
the proceeds of the sale of oil.

In turn, Hunt obligated himself to "drill all wells necessary
on the lands and premises hereinabove mentioned, to protect said lands and premises against drainage and waste on
account of producing offset wells on abutting property, unless in his opinion it would be unprofitable to drill a well
offsetting a producing well on abutting property, in which
event he agrees to abandon such lease, or part thereof, and
assign same to the party of the first part, if demanded . . .
provided that any part of such lease so abandoned shall
consist of not less than five (5) acres."

Two days after the contract-signing it became public
knowledge that a rich core had been taken from the Deep
Rock well. The price of Rusk County leases shot skyward.
Joiner, it was reported, was offered $3,500,000 for his hold-
ings by a major oil company which had not learned of his
deal with Hunt.

And the Deep Rock well came roaring in on December 13,
flowing 3,000 barrels per day of 40.5-gravity oil. This was a
real oil well, the kind Dad Joiner had hoped for, the kind all
wildcatters hope for.

On December 16 Hunt brought in his south-lease well. It
was a small producer, as Hunt had expected, making less
than 100 barrels per day. And Stroube and Stroube brought

in another small producer to the northeast which, like the Daisy Bradford 3, flowed by heads. Weeks earlier the well had almost been lost when floating gas ignited and flames swept the derrick.

On December 20 Hunt hooked his well, the Daisy Bradford 3 and the Deep Rock well to his pipeline and pumped the oil to the loading rack at the railroad. The first oil to leave Rusk County went to Sinclair's Houston refinery in thirteen tank cars carrying 10,000 gallons each. Hunt's Panola Pipeline Company was in business.

A week later he had a competitor. Inland Waterways Pipeline Company began to work on a four-inch line to the railroad where it was building a ten-car loading rack. To get a start in the field, Inland planned to pump oil from the Deep Rock well that Hunt's line couldn't handle.

A genuine boom was in the making.

chapter nine

THE GREAT
TREASURE HUNT

The flow of oil from the Deep Rock well had awakened the major oil companies at last, but most of them still refused to get out of bed. They did, however, cast a more speculative eye on Ed Bateman's Lou Della Crim 1, drilling near Kilgore.

And Bateman was encountering the usual difficulties that beset the poor-boy wildcatter. He was having little luck in selling stock despite his too-colorful prose, but he was pushing the well down with guts and blarney. His stock sales had been hurt when it became public knowledge that a major oil company had made a seismographic survey of Gregg County and northern Rusk County, including the Crim–Laird–Peterson land, only two months before the well was spudded in—and had reported "it is improbable that oil is present." Crim himself became so dubious of Bateman's ability to get the well down that he didn't buy a single share of stock.

Twice Bateman faltered. Each time he was encouraged by

a man who kept telling him, "Drill further, Ed, you're on the right track." He was Francis X. Bostick, a paleontologist and geologist for Southern Crude Oil Purchasing Company. Southern Crude had acquired some leases in northwestern Rusk County and had set up an office in Tyler. Bostick wanted a look at Bateman's well cuttings and cores, and Bateman needed all the help he could get.

Bateman was particularly depressed when his old rig sent the drill below 3,500 feet with no sign of the Woodbine. It was general knowledge that Ed Laster had cored the Woodbine at 3,486 feet on the Daisy Bradford 3. But Bostick assured Bateman he had not yet reached the cap rock—the Austin Chalk.

This information was almost identical to that given to Laster by the paleontologists who had examined cuttings from the Daisy Bradford 3—that he had not yet drilled to the Austin Chalk when, in fact, his bit was resting almost on the Woodbine. Bateman knew this story, and he was not immediately reassured by Bostick's assessment.

But Bostick knew the story also, and he thought he knew what had occurred. All of the formations in Rusk and Gregg Counties thinned out and ended against the Sabine Uplift, he calculated, and Laster had drilled through the Austin Chalk without being aware of it. Laster, he reasoned, had met so many difficulties in drilling that he easily could have missed seeing the chalky cuttings as they flowed from the borehole to the slush pit.

Bateman was drilling farther west from the Sabine Uplift than Laster had drilled. The formations would therefore be thicker, Bostick reasoned. Bateman, then, would have to drill deeper than Laster had drilled to reach the Austin Chalk. It would be quite thick—and so would the Woodbine.

"If there's oil around here, you'll make a good well," Bostick told Bateman.

Bateman continued to drill—and he reached the Austin Chalk late one evening. In three hours Checkbook Cain drilled exactly five inches. The "fish-tail" bits he had been using to cut quite easily through the other formations could not bite into the cap rock. The generator providing power for lighting broke down. Automobile lights were turned on to illuminate the rig floor. Cain lifted the bit off the cap rock. "We'll just have to rotate and circulate until we get a hardheaded roller bit," he said.

Word spread around Kilgore that Bateman was so broke he was out trying to borrow $86 to rent a roller bit. Bateman was far from flush, but the truth was that he and Ed Zilkey were in southwestern Rusk County borrowing a roller bit from H. L. Hunt. Hunt wanted Bateman to penetrate the cap rock because he wanted to know what was below it. He also loaned Bateman a core barrel; he wanted to know what coring would show.

The roller bit sped through the cap rock. On December 14, with only his associates, Crim and a curious Canadian stockholder on hand, Bateman ordered a core taken at 3,629 feet. The core barrel was opened on the rig floor. The bottom part of the core was oil-saturated Woodbine. Bateman leaped in the air, shouting in exultation—and saw half a dozen oil scouts approaching the rig. The scouts had got the word, as they almost always did, and the Bateman leap and shout told them that their trip to the rig was worthwhile.

Within three days the Texas Company, Magnolia Petroleum Company, Ohio Oil Company, Empire Gas and Fuel Company and Dixie Oil Company had acquired leases in the area, as had some small independent companies. If possible, the well was to be completed on December 28, a Sunday. Malcom Crim tried to persuade his mother, Lou Della Crim, to come watch the excitement. But Mrs. Crim went to church on Sundays, and this one was to be no exception. At 11:30 in the morning the well came in, flowing at a rate of more than 22,000 barrels of oil per day!

Crim, his shirt splattered with his oil, raced back to town to tell his mother about the great bonanza. Church was just letting out and he met her at the steps. He told her the news. "Come and see it! You've got to!"

Calmly, Mrs. Crim said, "Well, if you think it's safe, I'll come."

She went to the old farm and had her look, this serene and gracious woman who had told her father a half-century earlier that she would take the farm so that her beloved brothers might use his wealth and financial enterprises to build their futures.

She was the only calm person in Kilgore. By the following Wednesday the sleepy community of 700 had swollen almost incredibly to 10,000, and the greatest treasure hunt in American history was beginning.

Hardly anyone had noticed that on December 1 a wildcat test had been spudded in on the land Barney Skipper and Walter Lechner had leased some five miles northwest of Longview—just where Skipper had been pleading for someone to drill for almost twenty years. The site was thirteen miles north of where Bateman was then drilling the Lou Della Crim 1, and thus twenty-six miles north of the Daisy Bradford 3.

The location, on the farm of F. K. Lathrop, had not been selected by a geologist, but neither had it been picked at random. Lathrop, manager of the Kelly Plow Company office in Longview, owned a 400-acre farm almost in the heart of the Skipper–Lechner block, but on land Lechner had sold lease rights to John Farrell and his associates, W. A. Moncrief and Eddie Showers. Lechner, it will be recalled, had reserved the right to select the drill site.

Lechner had gone to Lathrop and told him he would have Farrell drill on the Lathrop farm if Lathrop would give Lechner and Skipper half of his one-eighth royalty. Lathrop had promptly accepted the proposition. At the farm, he and

Eddie Showers had chosen a drill site near the bank of a creek.

The Lathrop 1 was a wildcat test in every sense, but Farrell's was not a poor-boy operation. He had dealt interests in the well and his acreage to obtain excellent equipment, and the drilling was performed by competent crews. Nevertheless, there were occasional delays because there were no repair shops in the area; repair work had to be done in Shreveport or Dallas.

The completion of the Deep Rock well near the Daisy Bradford 3 on December 13, and completion of the great Lou Della Crim 1 near Kilgore on December 28 caused some attention to swing to the Lathrop 1. The Longview and Gregg County Chambers of Commerce, observing the enriching pandemonium in Henderson and Kilgore, announced a cash prize of $10,000 for the first well brought in within a ten-mile radius of Longview. The Lathrop 1, five miles from the city, would qualify if oil were found.

Back in Kilgore, Ed Bateman had solved his problem with the postal authorities. His great well was even better than he had claimed it would be in his colorful mailers. But because he still lacked cash when the well came in, he was unable to buy a proper Christmas tree, an assembly of valves and piping placed on a well to shut it in or control its flow. Checkbook Cain had assembled a makeshift Christmas tree of undersized pipes and fittings, but the pressure in the well was so powerful it threatened to blow the Christmas tree apart. Ed Zilkey had lashed down the weird device with chains and chain tongs in hopes of keeping it together.

Bateman had a well but he needed money to develop the rest of the 1,494-acre lease. He went out seeking loans, but found none. He decided to put his well and acreage on the market. He first sought buyers in Dallas and Shreveport, asking $1,500,000 cash and $600,000 in oil produced for his seven-eighths interest in the package. He failed in both places. But while he was in Shreveport, talking business with the Ohio Oil Company, a telephone call was received in his tiny Kilgore office by an aide, O. P. Douglas.

Douglas held his hand over the telephone while he inquired if anyone knew where Bateman was; a "guy named Pratt, who sounds like some kind of broker," wanted to talk to him. Ed Zilkey, the jack-of-all-trades, grabbed the telephone.

"Is this Wallace Pratt of Humble?" he asked.

"Yes."

"Ed's in Shreveport. I'll get him in touch with you."

Wallace Pratt was in charge of Humble's geological, scouting and land departments. He was widely considered to be one of the smartest and most daring geologists in the oil business. He had taken a look at the Lou Della Crim 1, and he had conferred with E. A. Wendlandt and E. J. McLellan, his regional geologists in Humble's Tyler office. Wendlandt, who had been chiefly responsible for maintaining Humble's interest in Rusk and Gregg Counties since 1927, had been enthusiastic about the Lou Della Crim 1 and the acreage around it. His enthusiasm had infected McLellan, a Scot who chose his words carefully and rendered his decisions in brief paragraphs.

Bateman met with Pratt and other members of the Humble hierarchy in the company's Houston office. He told them what he wanted—$1,500,000 in cash and $600,000 in oil produced—and he refused to budge from that figure during the long discussion.

Pratt wanted to deal. So did W. S. (Bill) Farish, Humble's president and one of the company founders. Other officials walked the floor and talked. Bateman was asking a big price. The Great Depression was in its second year without a sign of recovery around the corner.

L. T. (Slim) Barrow, now the company's chief geologist,

was in favor of the deal but subject to the well's satisfactory performance during a sustained test.

Bateman shook his head. "We've got a sorry Christmas tree on that well, Mr. Barrow. That well's not going to be tested as long as it belongs to me. It might get away from us and be ruined. If you want it tested, buy it and test it yourself."

Farish grinned, then asked Barrow, "What does Mac think of it?"—referring to McLellan, the conservative Scot.

"He says," replied Barrow, "that if we take it we'll never need to look back."

"Well, if Mac says that, what more do we need to know?" Farish asked around the board table.

On January 9, 1931, Humble paid Bateman's asking price. Humble dug deeply to pay it. But Farish and Pratt and the others were looking beyond the Depression to a day when the company would need great reserves to supply the future they dreamed of and were sure would come.

Bateman, money in hand and with more to come, left East Texas, never to return.

It will be recalled that when Walter and Leota Tucker went broke in 1927 and were helping Dad Joiner sink his two abandoned wells, Mrs. Tucker tried to sell her 305.6-acre farm near Kilgore, the only family asset. Now, with the completion of the Lou Della Crim 1, Sun Oil Company leased the farm for $100 per acre.

Mrs. Tucker was standing in the Overton State Bank, ready to deposit her $30,560 check, when H. L. Hunt came in to borrow $5,000. Hunt had completed his deal with Joiner, had drilled his south lease well and had begun operating his pipeline, but he had not yet made a profit on either enterprise. He needed the $5,000 to meet current expenses.

R. A. Motley, the bank president, turned him down, explaining that he had extended too much credit on projects that had failed so far to materialize.

Mrs. Tucker observed what was occurring. She had met Hunt. She knew about his deal with Joiner. She waited and stopped Hunt as he was about to leave the bank.

"I'll lend you the five thousand dollars," she said. "At eight percent interest."

Hunt smiled. "Just what I'm accustomed to paying, ma'am."

"With no security," she continued.

"It's a fine deal, ma'am," Hunt said.

Mrs. Tucker arranged for Hunt to get the $5,000, and in a few weeks he came to pay her back. She shook her head. "No, sir. I want to keep you on the hook. Just keep paying me my interest by the month until I let you know I want the principal."

Hunt gravely paid her $33.33. With the next month's payment he sent along a hundred-pound sack of paper-shell pecans, a delicacy he had learned Mrs. Tucker relished.

Was it two new oil fields or one huge one that reached 13 miles from the Daisy Bradford 3 to the Lou Della Crim 1? That it could be the latter was hardly conceivable, but there were some who broached the idea.

Writing in the *Oil and Gas Journal*, L. E. Bredberg, who had covered the completions of both discovery wells, painted this interesting picture: "Some believe that the intervening acreage will all prove fertile for production, while others are of the opinion that the Bateman producer is on another structure and is the harbinger of a separate and distinct pool. Whichever opinion proves correct does not make much material difference, for it now seems apparent that the Joiner pool will produce many large wells and will be a major pool, while the newly discovered area may prove likewise if it does not connect with the Joiner area."

As Bredberg had written, it did not "make much material difference." Independents and lease-hounds swarmed like locusts over the intervening countryside. And the major oil companies, who could have leased Rusk County in its entirety at any time in the past for a few dollars and a promise to drill, were now paying from $1,000 to $5,000 per acre for leases. Humble, with the knowledge of the area accumulated by its geologists to guide it, was in the leasing van.

But only Amerada and Shell had acquired leases in the vicinity of the Lathrop 1 test northwest of Longview in Gregg County where the bit had been steadily grinding downward since December 1.

On the night of January 12, 1931, the Woodbine was found and cored at 3,574 feet in the Lathrop 1. Before dawn the word had spread and by noon lease-hounds were out in force, paying more than $1,000 per acre for leases as far away as ten miles from the well. Sun, Tidewater, Stanolind and Simms were in the forefront of the larger companies, but as before in southwestern Rusk County with the Daisy Bradford 3, and then in northwestern Rusk County with the Lou Della Crim 1, the independents and individuals were the more aggressive leasers and less demanding in establishing validity of titles.

Dry Hole Byrd, the young wildcatter who had supplied Dad Joiner with good equipment to complete the Daisy Bradford 3, hit Longview like a duck on a June bug. He was flush; he and his partner, Jack Frost, had sold Humble a lease block of 7,000 acres and a one-fourth interest in another block of 2,800 acres on land laying between Kilgore and Longview.

Now he rented office space in the Gregg Hotel. He secretly bought out the Gregg Abstract Company operated by Hall Wood. He tied up, under contract, all five notaries public who were active in the county. He arranged with county officials to set up small temporary office buildings on the courthouse grounds. He had pinpointed seventy-two areas in the county where he calculated lease action would be at its hottest. He raced to Dallas in his Pierce-Arrow and hired seventy-two girl typists; he wanted each girl to concentrate on a separate volume of the seventy-two volumes of county abstracts which covered his seventy-two areas of interest.

As soon as the courthouse opened in the morning, the girls would rush into the county clerk's office and each would get the volume to which she had been assigned. This gave Byrd first claim on the books. With this claim, his contract with the notaries public and ownership of the Gregg Abstract Company, practically every lease transaction—from title work to consummation—had to pass through hands controlled by Byrd.

To a protestor, Byrd explained: "I'm not stopping anyone from getting their titles cleared; I'm just getting mine cleared first."

He was able to guarantee same-day-service to almost all the persons he was dealing with on leases. His office in the Gregg Hotel lobby had a window that opened on an alley. Byrd's employees did a lot of business through the window —taking in documents, paying out cash and checks. When Byrd found acreage he wanted, he could have it checked out by evening and he was able to pay the lease seller immediately.

It was almost as good as owning an oil well.

The completion of the Lathrop 1 was being awaited by the oil industry and anyone who calculated to make a dollar from it. The Longview newspaper carried advertisements, paid for by the chambers of commerce, inviting one and all to witness the completion on Monday, January 26, 1931, "between 11 and 12 o'clock."

More than 18,000 persons had gathered at the drill site by the appointed time. Some had been there since dawn; some had been "sitting up" with the well for days. There was a sea of automobiles stretched out across the broad acres, and the license plates said the automobile owners came from almost every state in the Union. Reporters and photographers from the state's newspapers, large and small, had converged on the site.

The well was brought in under control at 1:10 P.M., flowing 320 barrels of 39.6-gravity oil in the first hour through three-inch tubing and a half-inch choke valve. Had it been allowed to flow wide open, it could have made 20,000 barrels per day!

The well, the Longview newspaper said, "brings oil to Gregg County for the first time. This opens up a vast territory heretofore unproven, but believed to have been capable of production for some time.

"Being a 15-mile extension of the Kilgore field which is now experiencing frenzied activity, the Longview well opens a new field that may develop into one of the greatest in the entire United States, in extent if not in production."

The most immediate beneficiaries of the Lathrop 1 were John Farrell and his associates W. A. Moncrief and Eddie Showers. On February 7, they sold their interest in the well and acreage to the Yount–Lee Oil Company of Beaumont for $3,270,000!

The Longview and Gregg County Chambers of Commerce came up with $8,000 of the promised $10,000 for the first well in a five-mile radius of the city, and Farrell gave the prize money to the drilling crew.

Barney Skipper, Walter Lechner and Ray Hubbard were sitting on 4,300 acres of some of the most valuable real estate in America. Skipper had fulfilled the promise Old Man Bill had made the farmers in the area when the century was young. Now he, Lechner and Hubbard set about making themselves millionaires.

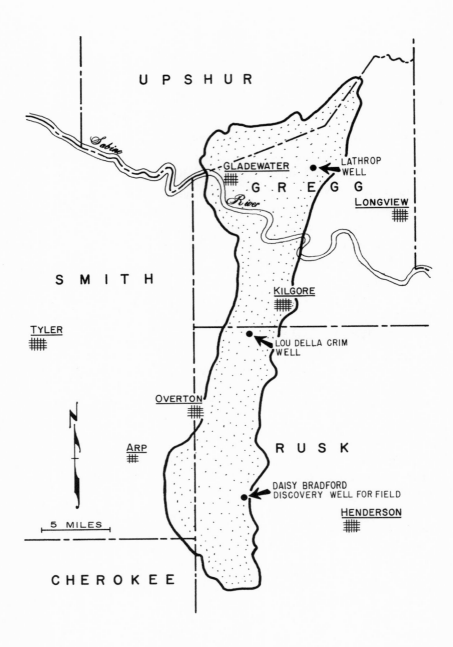

Outline of the giant East Texas field showing location of the Daisy Bradford discovery well for the field and the two key extension wells, the Lou Della Crim and the Lathrop.

chapter ten

THE BLACK GIANT

I t *was* one field. The Daisy Bradford 3 and the Lou Della Crim 1 were producing from the same field—and so was the Lathrop 1! Derricks marched across the intervening miles, then marched both north and south and to the west. They studded the landscape in five counties: Rusk, Gregg, Upshur, Smith and Cherokee. Oil was spouting from a reservoir forty-five miles long in a north–south direction and five to twelve miles wide from east to west. In all, it covered more than 140,000 acres!

It was the greatest oil field ever seen on the North American continent!

How then, did the scientists from the oil companies, large and small, miss it? And how did Doc Lloyd know it was there?

"Do not be discouraged," Lloyd had written Joiner back in 1927, "for you will surely be successful in discovering one of the largest oil fields in the world." And with astonishing accuracy he had forecast: "You are certain to make a well in the Woodbine Sand at 3,550 feet."

His report was laced with inaccuracies; fabrications, perhaps, would be the better word in some instances. But the overwhelming fact was that he had said Joiner would find oil, that he would find a great accumulation of it, and he had pinpointed almost exactly where it would be found.

As H. L. Hunt had calculated, the Daisy Bradford 3 had been drilled too far to the east to be a good well. Joiner had drilled on Miss Daisy's farm because he had made a profitable deal with her. But Lloyd had begged Joiner to sink his wildcat well some two miles to the west of the Bradford farm. Had Joiner done so, he would have drilled in what oilmen later called the "fairway," a lane running the length of the field, almost midway between the east and west boundaries, where the Woodbine was thickest and most highly saturated with oil.

Lloyd received little or no credit, however, from the oil industry generally and geologists in particular. "Luck," the red-faced scientists said. "He's just a veterinarian," the rest of the industry said.

Only a few remembered that it had been Lloyd who had told Joiner where to drill in Oklahoma—and that Joiner would have discovered the Cement and Seminole fields had he been able to drill deeper. Lloyd had other discoveries to flaunt had he chosen to do so.

But he did little to soothe his critics. He was beefy and domineering, and he explained nothing. He ruffled feathers when he spudded in a well of his own while a brass band blared and entertainers—including a son, Tex—put on a lively show for more than 1,000 witnesses. Another son, Douglas Lloyd, fiercely devoted to his father, roughnecked on the well. He had helped Lloyd map southwestern Rusk County in 1927, and he was ready to defend his father's geological talents in all quarters. He supplied the background for his father's success when Doc would not do it for himself.

The young Lloyd admitted to the promotions, to the medicine show, but he told of his father's studies, his oil-field finds. He was tough enough to discourage too intimate questioning.

On one occasion, Lloyd himself told a small group in Boyd's Cafe in Henderson: "I'm not a professional geologist in that I didn't study prescribed courses in a recognized school to acquire a degree in geology. But I've studied the earth more, and know more about it, than any professional geologist now alive will ever know."

Lloyd stayed in Rusk County but a short while after the boom commenced, though he undoubtedly would have become a millionaire had he not departed. His photograph was taken standing beside Dad Joiner and Miss Daisy on the site of the Daisy Bradford 3. This picture, and others in which he appeared, were printed in newspapers across the country. Women in far places who did not recognize the name were quick to recognize the face and physique. The stationmaster at Overton remarked that every time a train stopped a different woman got off, sometimes with one or more children, and began inquiring about Doc Lloyd. "They're ganging up on him," the stationmaster told the crowd in the bank.

In any event, Lloyd quietly left the field. Perhaps there were other fields he wanted to find.

The field was so huge that it could not be named for any specific locality, as were most fields in Texas and elsewhere. It had to be named after an entire region: officially it was called the East Texas field. The more romantic called it the Black Giant.

The sheer size of the pool brought the greatest discomfiture to the geologists who had said it wasn't there. Some geologists had wanted to devote more study time to the area than their companies would permit, and a few of them grumbled that it was the geophysicists, not they, who had been at fault.

It was true that oil companies in the late 1920's were enamored of the refraction seismograph, the torsion balance, and later the reflection seismograph. They had reason to rely on the instruments and their operators: oil was being found in abundance by seismic crews. On the other hand, when seismic tests failed to locate structures favorable to trapping oil in a certain area, companies were sometimes prone to dismiss it even though geologists thought it promising.

The first seismic instrument used in Texas, in 1922, was the torsion balance. This tool was used for measuring the pull of gravity in the hope that it would help locate salt domes and other formations having a density differing from that of surrounding rocks.

In 1924 the Marland Oil Company began using the refraction seismograph in Texas. It was expected to help determine the depth and character of formations by recording the speed and quality of shock waves sent through them and refracted to the surface. Other companies quickly followed Marland's lead. Some companies began using both instruments.

The success of the instruments was beyond all expectations. Some of the best and largest fields of the period on the Gulf Coast, for example, were drilled as a result of seismic testing. In Sugarland, near Houston, Humble geophysicists found a salt dome buried more than 3,300 feet below the earth's surface—and the drill found oil.

But the East Texas field had been traversed countless times by seismic crews who found none of the familiar indications of oil. They had found none because there were none, Doc Lloyd's report to the contrary. Not only was East Texas the largest and richest pool on the continent, it was unique. It had been a puzzle because it was mystifyingly simple, as geologists learned when its composition had been established. Arriving at the truth was like taking a refresher course in basic geology:

The earth is several billion years old, and consists of various layers of material. Man lives on the surface of the earth's crust, which might be compared with the peel of an orange. The crust consists of the great variety of observable rocks. It is thickest beneath mountain ranges, where it may reach depths of forty miles. It is thinnest at ocean bottoms—sometimes less than two miles deep.

Beneath the crust is a layer called the mantle, a zone of tremendous heat and pressure. It is about 1,800 miles thick and contains very heavy and hard rock. This rock, however, will bend, distort and even flow as a plastic when subjected to extreme pressures. This, in turn, will affect the crust above it: the most severe earthquakes are believed to originate, or be generated, in the mantle.

Below the mantle is a layer believed to be of liquid iron and nickel. It is about 1,300 miles thick. It is called the earth's outer core. At the center of the earth is a solid ball of iron and nickel about 1,600 miles in diameter. This is the inner core.

The planet, then, is comprised of diverse layers of materials, heaviest at the center and lightest at the surface. Tremendous pressures and temperatures occur deep within the crust, and even greater heat and pressure exist in the mantle. The crust is constantly being shifted through erosional processes, and the resulting shift in weights of the crust is exerted on the mantle. Also, the planet is constantly subjected to gravitational and centrifugal force as it rotates around the sun and as the moon rotates around the earth.

These factors produce internal pressures which cause rocks to swell, break and move. This causes the earth's crust to bend—to rise here and create a mountain range, or to fall there so that lowlands or islands become submerged beneath a sea. Wherever a sea occurs in this manner, sediments are brought to it by rivers and streams from adjacent eroding land masses. Thus, material is transported from one area to another and the weight of the crust is continuously being shifted.

As the sea encroaches on a land area it deposits extensive layers of sand, mud and limestone over the previously dry land. Sands usually are deposited nearest shore, muds (or shales) further seaward, and limestone yet further seaward simply because the heavier particles of sediments—sands in this instance—settle out first, or nearest shore. When the sea retreats away from the land these deposits are exposed to the air, become part of the land, and are subjected to erosion, just as present land surfaces are being eroded by running water and the weather.

If the sea surrounds an island, it will deposit sediments around the island. The island will receive no such sediments and will in all probability lose material to the sea through erosion. If the island, through earth movement, begins to rise, the sea will retreat from its previous shores and the island will grow by encompassing land which previously was sea bottom. At the same time erosion of the uplifted, larger island will accelerate since the speed of running water on the surface will be increased as the island rises.

On the other hand, if the island begins to sink, the sea accordingly will encroach on the land and erosion of the island's surface will slow down because of its lowered elevation. When the island sinks below the surface of the sea, it will be eroded by wave action. When it sinks sufficiently to lie below wave action it will begin to receive sea deposits, just as the surrounding sea bottom does.

Eons and eons of geologic time ago, an extensive sea covered a region which included the southeast half of the present state of Texas and the north half of Louisiana. A tremendous salt bed several thousand feet thick was deposited throughout large portions of the region, probably because of climatic conditions which changed the ratio of salt to sea water. The salt beds are believed to have been laid down during geologic periods called the Triassic and the Jurassic, roughly one hundred seventy-five and one hundred fifty million years ago.

Following this salt deposition, the sea fluctuated back and forth over the land, advancing landward as the land sank, retreating back toward the present Gulf of Mexico shoreline as the land rose. As it moved back and forth, it laid down layer after layer of varied sediments; thousands of feet of material—sand, shales (mud) and limestone—had been deposited over the salt beds by the end of a geologic time period called the Lower Cretaceous, some one hundred to one hundred twenty million years ago.

At this time a large area that had been part of the sea bottom slowly began to bulge upward in what is now northeast Texas and northwest Louisiana. This rise probably was caused by movement in underlying rocks of the crust and mantle. As the sea floor rose, the shallowing sea deposited a widespread layer of sand several hundred feet thick over limestone beds which previously had been deposited when the water was deeper. This sand would become the precious Woodbine.

After the sand layer was deposited, the upward bulging of the area apparently was interrupted by a relatively short period of sinking. This caused temporary deepening of the sea and resulted in deposition of a thin layer of mud—which would become the Eagle Ford shale—over the thick layer of Woodbine.

Again the region rose so that a portion of the sea floor reached above sea level and formed a broad island. The island's higher elevations lay along and just to the east of the present Texas–Louisiana border. In Texas, therefore, the sea shallowed eastward and was deeper to the west; its Texas shoreline lay along the west side of the island, which now was being eroded by running streams and weather.

As the island continued to rise and grow in area, the sea retreated farther westward in Texas. Consequently the is-

land suffered still greater erosion, which removed all Woodbine sand, the Eagle Ford shale above it, and substantial
portions of the underlying limestones from the areas of
higher elevation. Streams carried these materials into the
retreating sea and redeposited them over the Woodbine
where it had not been exposed to erosion.

As the sea lapped against the west side of the island it
moved landward or retreated in accordance with minor
fluctuations in the island's vertical movement. As a result,
the Woodbine on the beach and on the shallow sea bottom
was reworked by wave action and redistributed along the
shore and in the shallow sea waters. Eventually the island,
the surface of which was of weathered, eroded limestone,
was flanked by a wedge-shaped body of sand.

This wedge of sand was thickest in the sea to the west and
disappeared to the east on the west shore of the island. It
stretched in a north–south direction for the total length of
the island, and its eastern edge formed a sandy beach on the
weathered west side of the isle.With its limestone surface
and sandy shore, the island remained above sea level for a
long period. It was the time of the great dinosaurs, huge
flying reptiles and land-bound reptiles, including crocodiles.
Small mammals scurried to keep out of the reach of these
creatures and flesh-eating birds. The climate was probably
warm and humid. Flowering shrubs, herbs, and trees like
the oak and elm possibly grew on the island. Sharks lurked
in the adjacent sea.

Abruptly, the island sank beneath the water. And the sea
covered it and the surrounding sea bottom with a layer of
chalky material. The sea remained there for millions of
years, and as the crust continued to sink, the former island
and the entire Gulf Coast Province were buried under thousands of feet of ocean sediments.

The chalky material which covered the island and the
surrounding sea bottom would become known as the Austin
Chalk. The wedge of Woodbine, which made up the beaches

and the adjacent sea bottom, thus became the meat in a figurative sandwich. The bottom slice of bread was the limestone on which it had been deposited and which formed the weathered surface of the island. The top bread slice, of course, was the Austin Chalk, which had been deposited over the entire region, including on top of the island. Where the sandy beach terminated, the Austin Chalk rested on the island's limestone surface; there was no intervening Woodbine.

Again the entire region was uplifted, and a great retreat of the sea toward the present site of the Gulf of Mexico took place. Although the water oscillated back and forth over the land during this retreat, recessions of the water to the south predominated over its advances to the north. Thus the present land areas of the Gulf Coast Province emerged from the sea which is now the Gulf of Mexico.

And again the buried island area was uplifted, probably by the same type of earth forces which originally caused it to rise out of the sea. This time the area was a bulge, not an island. It would become known as the Sabine Uplift.

Oil was formed from the remains of plants and animals that grew and lived in the sea or on the land mass. These remains eventually became a part of the sediments deposited in the water.

As the sediments were buried under additional deposits, heat and pressure increases induced changes that converted the organic remains to petroleum. With additional burial, more heat and pressure forced the oil out of the sediments in which it was formed—usually shales—into water-bearing porous rocks, such as sands like the Woodbine, or porous limestones. There, mixed with salt water, oil sought higher elevations, just as it will float above the water in a container in which oil and water are placed.

When the Sabine Uplift became an elevated area for the second time it was pushed up from below so that buried layers of sediments sloped away from the highest parts of the uplifted region. In Texas, therefore, the uplifted area, including that occupied by the Woodbine sand wedge, sloped towards the west, just as it had after the original uplift formed the old island. This caused oil which had entered the Woodbine sand from source areas to the west to migrate eastward through the Woodbine sand, seeking higher elevations. The sand not only contained oil, it held *much* more salt water. The oil therefore floated *through* salt water and *around* sand grains as it made its way eastward until its route was cut off. It had reached the old shoreline where the Woodbine played out against the old island's limestone surface—and was capped by the overlying Austin Chalk. Because they were both impermeable—so dense they prevented the oil's movement—the Austin Chalk above and the limestone below proved to be a barrier to further oil migration.

Oil began to fill up the easternmost edges of the sand wedge. It spread north and south along the old shoreline, then began to fill the thicker layers to the west until it had accumulated in an area of about 140,000 acres.

But an area like the west flank of the Sabine Uplift—where only an unbroken west dip occurred in buried sediments and no sign of structure could be recognized—was no place for scientists or anyone else to look for petroleum.

So Dad Joiner looked—and tapped an ocean of oil.

No sooner had the three key wells in the field started flowing than "if" stories began circulating in East Texas.

If Joiner had drilled a quarter of a mile to the east he would have missed the field. True, but had he drilled on a site selected by Doc Lloyd—two miles to the west—he would have brought in a much larger well because there the Woodbine was much thicker.

If Humble had drilled, as planned, on its London block in

the summer of 1930, it would have discovered the field. But Humble didn't drill; indeed, Humble bought its first oil in East Texas with the purchase of Ed Bateman's holdings in the Lou Della Crim 1.

But the most fascinating "if" story concerned the drilling of a dry hole near Kilgore way back in 1915. It was drilled by Roxana Petroleum Company, a Shell subsidiary, on acreage that had been assembled by Snowden Brothers of Indianapolis. Snowden Brothers had hired the geological firm of Fohs and Gardner to study the area and select a drill site. Fohs and Gardner were F. Julius Fohs and James H. Gardner, two young geologists with an office in Tulsa.

The story went that William Anton Jesus Maria Van Waterschoot van der Gracht, a renowned Dutch geologist high in Shell management circles, arbitrarily moved the location for the test well a mile and a quarter east of the site selected by Fohs and Gardner. The test was dry. By moving the location that mile and a quarter east, the story said, van der Gracht had missed finding the East Texas field.

The story was widely accepted, and Fohs, by then a well-known and successful geologist, did nothing to discredit it. In fact, he often told the story himself.

But Gardner said the story was not true—and it was he, not Fohs, who had studied the area and had selected the drill site. "It is a tall tale without foundation in fact," Gardner said in a written statement.

Gardner studied a large portion of East Texas and, like several geologists who came after him, became intrigued with the Sabine Uplift. The Kilgore area was particularly to his liking because he thought he detected some sort of structural folding. Nothing was known about either the presence or the absence of the Woodbine in 1915. "We were shooting in the dark as to what the true structure and sands might be," Gardner wrote.

Jack Shannon, Snowden Brothers' manager, employed Dr. E. H. Hamilton to block up about 15,000 acres in the

Kilgore locality. Dr. Hamilton was a beloved physician who drove around the countryside in a red Ford runabout. He quickly assembled the block, and Snowden Brothers gave Roxana a half interest to drill the test well.

The site was examined by van der Gracht and his assistant, R. A. Conkling—and they accepted Gardner's judgment. The well was sunk by Carl Lemon, a drilling contractor, and Burton Hartley, a Roxana geologist, "sat on the well" to inspect cuttings. The test was two and three-quarter miles southeast of Kilgore on the Fambrough farm—and it was one and three-quarter miles east of the limit of the field. It was abandoned at 3,482 feet.

Gardner, however, offered another story which may have sired the first one. Shannon, Snowden Brothers' manager, supervised the unloading of the drilling rig during a heavy rain. Roads to the drill site were impassable. He asked the Roxana supervisor to let him rig up and drill at a handy spot on the west side of Kilgore. The Roxana supervisor refused. If the well had been drilled at Shannon's handy spot, the drill would have found the Woodbine. Gardner didn't hear this story, however, until after the field was discovered.

In burying the story concerning Fohs–Gardner and the renowned Dutchman, or trying to, Gardner wrote: *"In thinking back over the history of this undertaking, it occurs to me that if this test had been placed at a point that would have accidentally discovered the East Texas field, my personal reputation for locating oil fields would have been magnified out of all proportion to its justification."*

But the "if" story was too good to die; it was told wherever oilmen gathered.

2

THE EXPLOITERS

chapter eleven

BOOM AND BEDLAM

While factories stood idle across the land and heartsick, hopeless Americans stood in bread lines from New York to Los Angeles, East Texas, in the spring of 1931, was caught up in the most frenzied boom in the nation's history. It was the California gold rush, the Klondike, the Oklahoma land rush and the wildest of past oil booms rolled into one. The daring, the resourceful and the unscrupulous had caught the scent of oil and money in the spring air, and they swooped down on the five oil counties like the cavalry of Genghis Khan.

They overran Henderson and Overton, Tyler and Gladewater, Kilgore and Longview, and all the crossroad settlements in between. They invaded villages that lay outside the immediate oil province—communities like Arp, Troup and Turnertown. They erected jerry-built towns and tore them down. Joinerville, the first boom settlement, almost burned to the ground on the night of March 4 . . . and was forgotten.

Oilmen were the first wave—wildcatters, roughnecks, roustabouts and lease-hounds working for themselves or for companies large and small. They were greeted with cries of joy from the East Texans, who had lived on the edge of starvation for almost five years. For the oilman brought money or was capable of making it, and every farmer and merchant wanted all he could get of it. Farmers ignored their fields and livestock to devote their time to leasing out their land parcel by parcel. These new lease-hounds were not trying to assemble blocks of acreage, as Joiner, Crim, Skipper and other wildcatters and major oil companies had done. If a piece of land was large enough to hold a derrick and a drilling rig, the newcomers were ready to pay the price. So farmers leased their acreage in segments for higher and higher prices, and merchants tore down their buildings to clear space for drill sites.

There was no semblance of order. There was little time to eat or sleep, and few places to do either. Everything was done in a hurry. A flat tire on the way to a farm could cost an oilman a lease—and a fortune.

The disorder appeared benign because it produced money and jobs to make more. Experienced roughnecks working ten or twelve hours daily on a drilling rig could make $5 or $6, drillers $10 or $12. And there was God's plenty of work for drillers and roughnecks; new wells were spudded in as fast as derricks could be erected. There were no dry holes once the limits of the field had been defined; every well was a producer and every day was payday.

And the landowners flourished. In 1929 and early 1930, Sam Ross, Kilgore's druggist, had tried without success to lease his 8,000 acres for $1.50 an acre. Now he leased it in parcels, getting $1,800 to $3,000 an acre! Farmers bought new automobiles, built new homes, painted old ones.

This prosperity came about primarily because the major oil companies had not believed there was oil in East Texas. Had they thought otherwise, they would have leased up large blocks at rock-bottom prices and drilled at their own pace. But now East Texas was cut up in tiny segments like a giant jigsaw puzzle, and most of the leaseholders were individual operators and small independent companies who sent their drill bits to the Woodbine as fast as they could spin them. The majors were forced to scramble for high-priced leases and drill at the other man's speed.

And speed they did. It had taken Ed Laster sixteen months to complete the Daisy Bradford 3. But now professional crews with good equipment could move in and complete a well in less than a month. Many wells were being completed within fifteen days, and Sinclair completed its Holland 1 in nine days.

Not even Dad Joiner knew how much it had cost to drill the Daisy Bradford 3, but now wells were being completed for $25,000 and less. Much of the drilling was done by contractors at $3.50 per foot with the contractor furnishing drilling water and digging the slush pit; the oil operator supplied the fuel oil for the boilers at 75¢ a barrel.

In its March 6 issue, *The Oil Weekly* broke down drilling costs this way: Drilling to 3,600 feet, $12,000; wooden derrick, $1,000; fuel oil at 75¢ a barrel for 1,500 barrels, $1,125; casing, $6,167; cementing, $475. The total, $21,367. An additional $4,500 for a fuel tank and lines and three 1,000-barrel tanks to hold the oil brought the entire cost to $25,867.

An expense common in other oil fields did not at first disturb the East Texas operators. The huge walking-beam pumps, which cost more than $5,000 when installed and ready for operation, were not needed; oil flowed from the wells under gas and water pressure. Yet many operators, and even major companies, shipped pumps into East Texas before they learned better. The pumps rusted in the mud for months before they were hauled out for use elsewhere.

Kilgore became the heart of the boom. And the derricks, side by side, moved on Kilgore—and Kilgore stepped aside. Buildings were moved or razed to make way for drilling.

Lou Della Crim, on whose farm the great gusher was drilled, could watch from the window of her town house and see three wells being sunk in her backyard. And when those came in as producers, three wells were drilled in her front yard! Kilgore became a jungle of derricks in which the homes and business houses appeared almost incidental. From the edge of town a man could travel for miles without touching the ground by leaping from derrick floor to derrick floor!

The Crim family owned six business lots in Kilgore that were covered with buildings. They were tempted to move the buildings or simply tear them down, but they finally decided to cut off the rear twenty-five feet of each. Six wells were drilled behind the six truncated buildings, each well standing on a twenty-five-square-foot plot! The derrick legs were touching.

A bank was torn down to make way for two oil wells. Others followed the Crim example and cut off portions of buildings in order to drill. Kilgore was a one-street town, and the business houses originally faced the railroad station and trackage. Soon new arrivals at the depot could see only oil wells from the train windows; the business houses now faced the alley, which had become the main street. There were forty-four wells on what would have been one city block!

The heavy rain that fell for days immediately after the Daisy Bradford 3 was brought in had turned East Texas into a swamp. And the ground never seemed to dry to hardness before another rainstorm struck. The streets of every boom town in the area were impassable by automobile. Laden with equipment, oil-field trucks sunk to the fenders in mud. Mules and oxen pulled them free and finally replaced them. The whip's song and the drivers' colorful profanity were heard around the clock. Blacksmiths came out of an enforced retirement.

Pipelines could not be constructed quickly enough to move the oil from the great field, and rumors rapidly spread that the majors were stalling construction in order to "break the independents." It was argued that the majors were slow to build pipelines in the hope that independents with no market for their oil would sell out cheaply.

Railroads found themselves trying to accomplish the impossible. Suddenly they were swamped with orders for hauling heavy boilers, lumber, steel, pipe and drilling tools. Just as suddenly the demand for tank cars soared, and long lines of them waited for days to be pulled to the loading racks. Tracks within the field became so cluttered that passenger trains were inched through, causing general disruption of schedules.

Truck lines tried to take up some of the slack, but the roads were so bad that the heavy vehicles often could not reach their discharge points until days after their estimated times of arrival. Another problem the truck-line operators met was described in a March issue of the *Oil and Gas Journal.* The article said in part: "Many contractors have found that the State Highway Department of Texas has been the initial cost for entering this field, for daily their trucks are being stopped on the state highways by state motorcycle patrolmen who weigh the loads and, in many cases, assess a heavy fine or tie up the trucks in the various towns along the way until the owner can purchase the necessary permits or pay off the fines."

Many patrolmen had a wrong conception of their duties, the article said, and made "insatiable demands" on the truckers. As much as $700 in one day had been assessed on trucks hauling equipment from New Mexico and West Texas fields, the article said, and one operator said he believed more than $1,000 per day was being collected in fines from truck owners in many of the towns in East Central Texas.

Both the state and individual patrolmen were responsible for this exploitation of the truckers. With its treasury bare, the state wanted all possible revenue from fees. Some patrolmen were in league with small-town magistrates, who demanded exorbitant fines which they shared with the patrolmen. This was a racket much like the speed trap of modern times.

If the State Highway Department was responsible for the initial cost of entering the field, a second payment often had to be made to a landowner or someone to whom he had leased the surface rights to his acreage. Some farmers charged $1 per vehicle for the right to move across their land. Others made roads of logs in swampy areas and charged a toll for their use. Some farmers were accused of plowing up local roads to force use of their toll roads.

The practice became so widespread that the Longview newspaper published an appeal to reason from F. K. Lathrop, on whose land the third major well in the field had been drilled. Said the newspaper:

> Attention is called generally to farmers in the Lathrop area urging them not to fall victims to "oil-field boomers" who are making a practice of paying a small rental on the surface rights of farms and then forcing the operating companies to pay exorbitant fees to travel over the property . . .
>
> F. K. Lathrop . . . asks all property holders to refuse to sell the surface rights on their farms, stating it will retard development and cause them to be losers in the end . . .
>
> Should the farmers continue to sell their surface rights, drilling programs in the Lathrop area will be paralyzed shortly. The paying of the fee is very annoying and must be stopped, according to Mr. Lathrop, if the oil area is to be developed. The operating companies are reported as saying they will not tolerate such practice and had rather delay drilling than adhere to the requests for such rates . . .
>
> In one particular instance, a "boomer" purchased for a very small amount the surface rights on a tract over

which several hundred vehicles pass daily, and in turn attempted to charge one dollar for every trip.

Lathrop and the newspaper were ignored.

Behind the oilmen came the opportunity seekers: unemployed workers from virtually every state, thieves and gunmen, con men and gamblers, pimps and brigades of pajama-clad prostitutes from Dallas, Fort Worth, Houston and Galveston. Gay beach pajamas were both the hooker's street attire and her trademark.

Tent cities sprung up along roadways, near railroad rights of way, and around drill-site clusters. Four miles east of Kilgore, in an area called Newton Flats, a settlement mushroomed over three acres. It was devoted to vice—and it did not cater to the effete. Women, booze and every gambling game known to man were offered to rich and poor alike; a boomer could "shoot a dime" in one dice game while at the adjoining layout a more prosperous citizen could toss a thousand-dollar bill on a single spin of the roulette wheel.

Four miles in another direction from Kilgore was Pistol Hill, a hangout for thieves and stickup artists. With Pistol Hill as their headquarters, these outlaws spread out on forays all over East Texas.

A New Orleans reporter wrote that the "highways and byways of this new oil center are lined with taverns called 'honky-tonks,' and the music emanating from their open windows is enough to drown out the steady roar of the working machinery." He was quick to point out that the music was not jazz, "but fiddle and guitar music, mostly, with the singer lamenting over a lost love."

The honky-tonks were drinking and dancing spots. Few served meals of any kind. Even Mattie's Ballroom, a huge, glittering dance hall midway between Kilgore and Longview, provided only sandwiches for the tireless drinkers and dancers who crowded the place every night of the week. But Mattie Castleberry, the hostess, imported seven-piece orchestras from as far away as Austin and San Antonio to supply music for her weekly costume balls. In Mattie's Ballroom, many a future millionaire danced with an adventuresome girl who later shared his life and millions.

Another honky-tonk which was crowded nightly was a big, rustic oasis near Gladewater. It was raided often, to the delight of the patrons, many of whom continued dancing while peace officers smashed bottles and overturned tables in their zeal. The operators, like the operators of other honky-tonks, placidly accepted arrest and paid fines as part of the expense of doing business. The honky-tonks were not bawdy houses, nor did they cater to criminals. The fights which erupted in them generally were caused by drinking, jealousy and the exuberance of young men who naturally enjoyed testing their strength and toughness.

Every boom town had more than its share of vice, and little or nothing was done to curb it in the first months of the boom. The lone constable or the two-man police force in the communities simply was overwhelmed by the influx. Some tried to fight crime and were helpless. Others took payoffs and waxed rich.

But East Texas was also a haven for the unemployed who had the courage and desire to invade it. Skilled carpenters who had been making $1 a day—if they found work at all at home—built $75 houses in one day flat, and the ring of their hammers joined with the clanging of drilling machinery to produce a maddening cacophony. Sawmills were reopened and new ones built. Tall pines crashed to earth one day and were a cafe or derrick the next.

The workers lived in tents until they could afford a one-room house—and some stayed on in tents beyond that time. Those with families brought their bedding and cooking utensils with them. Some lived in tents and conducted their business in adjoining tents. There were tent laundries and dry-cleaning plants, grocery stores and drugstores, clothing stores and cafes. Blacksmiths labored under tent roofs. And young doctors who had fled the bankrupt cities of the East and South officed in tents and delivered babies in tents. Such a style of living and working was accepted as a badge of pride and independence. It was the America of the pioneers.

In Kilgore, Guy J. Stampff and C. McPhall laid out a 125-acre site; within weeks it was filling up with residences, stores, rooming houses, hamburger stands and welding and machine shops. Oil-field supply companies erected giant warehouses and machine shops along the muddy roads in every direction. They employed hundreds of carpenters, bricklayers and other journeymen in the building trades.

In Longview and Gladewater the small hotels were overflowing and oilmen were invited into private residences. An eight-story office building was under construction in Longview, and two sixty-room apartment houses and seventy other dwellings were being erected. On a typical night in Longview's Gregg Hotel lease traders and other oilmen stood shoulder to shoulder in the lobby to conduct business.

Henderson's population had swollen to more than 9,000. The community of Arp became the shipping point for railroad shipments destined for the field in the area of the Daisy Bradford 3. The largest town in the five-county area before the boom, Tyler had served as East Texas headquarters for some oil companies since the discovery of the Van field. Now thirty-three companies established offices there, and almost all of the larger independent operators in the field set up land-leasing headquarters. Tyler had several office buildings and two large hotels, the Tyler and the Blackstone. The Blackstone added nine stories to accommodate the newcomers. Overton felt the impact of the boom, but not as much as the other communities. Existing buildings were inadequate to cope with the influx, but landowners wanted to lease, not sell, their property to supply firms and to other businesses seeking building sites. The companies wanted to buy, not lease, so they located in Henderson and Kilgore.

By the time the boom was nine months old, twelve schools were standing in the Kilgore area alone. More than 3,000 students were in attendance, 2,600 white, 460 black. Schools ran on full-day sessions, with 63 teachers at work. One school had more pupils than there had been people in Kilgore before the boom!

Although they became richer through more generous tithing, the churches were near casualties. Said the Longview newspaper:

> The churches, like every other part of communal life, have undergone radical changes, and not since the sixties and the early seventies have they met with such complicacy in undertakings and at the same time looked out on such bright and promising prospects for the future.
>
> The story is the same in all of the churches. Almost overnight recognized "pillars" of the churches were driven from their posts of service by conditions in business affairs which demanded all of their time and physical strength.
>
> A seemingly insurmountable conflict arose. Church activities were confined largely to Sunday. The church people caught in the maelstrom of boomtime confusion found themselves unwilfully working on Sunday, or if not at work, so worn and weary from the long strenuous days of the week past that they were compelled by physical laws to take the day for rest.
>
> A general change in the churches resulted. Pastors preached to empty benches. Teachers taught to empty Sunday school classrooms. Deacons didn't deac . . .

Negroes fell victim to the quick-buck artists. Life for them in East Texas in 1931 was little different from life in 1851. Nevertheless, many owned acreage granted them by their former masters. Like the Indians in Oklahoma, many of them sold mineral rights worth millions for enough money to buy a new automobile. Circumstances had made them ignorant and illiterate. Bewildered by the complexities of contracts, they would sell their one-eighth oil royalties in their acreage in the belief they were selling only a portion of it. "What does this 'royal tea' mean?" they sometimes asked their white friends.

Many whites were outraged at the treatment of the Negroes. Malcom Crim repeatedly made efforts to keep them from selling their royalties. On several successive Sundays, he mounted the pulpit at various Negro churches and warned the members not even to talk to lease-hounds. Some heeded him and became wealthy. Most ignored his advice and sold out. But Crim and his family were so beloved in the black community that many came to him for instructions on how to deal their acreage or royalties. When the Crims saw that their Negro friends were intent on selling, ignoring advice to the contrary, they oftimes bought the leases or royalties themselves at higher, fairer prices.

The Crim family's boom-time concern for the Negroes was in keeping with their past concern for all of their neighbors and customers when during depressed times they had extended credit from their general store far beyond reasonable limits. Another indication of their humanity came when the boom was first flowering. With their wealth assured by the Lou Della Crim 1 and their holdings around it, they tore up all the debts owed them and burned the scraps of paper. "We're wiping the slate clean," John Crim told the customers. "We're even with everybody." The Crims put on a sale and sold everything in the store from the ground up, then leased the land.

The Crims were not alone in maintaining their integrity. The boom inspired greed which prompted corruption in many East Texans, but far many more remained true to their fundamentalist traditions. Many a lease-hound and independent operator testified to the East Texan's rock-bottom honesty. Dry Hole Byrd, for one, had agreed to lease 300 acres near Kilgore from a farmer named J. W. Watson for $200 an acre before the Lou Della Crim 1 was brought in. The great well was brought in before Byrd could complete the deal, and Watson was offered $1,000 an acre by other oilmen. But Watson waited for Byrd to draw up the lease agreement and held true to Byrd's $200-per-acre offer because he had said he would.

Watson's land was productive, but he didn't live long enough to see it. While Byrd's crews were drilling the Watson 1, workers for another company began moving equipment across Watson's land to reach a nearby railroad right of way where a well was to be sunk. Leaving Watson on guard, Byrd rushed to Longview to seek an injunction against the trespassers. Watson sat on the levee bearing the trackage. Because of the drilling noises, he did not hear an approaching train and was killed on the spot. Watson's life had been hard—and death came just as his life was brightening.

There were similar tragedies, and in them the natives saw the hand of fate more often than blind chance. Guy Lewis had grown up on a farm in southwestern Rusk County, but had left at an early age to work in oil fields around the world. In Louisiana he had loaned drilling tools to H. L. Hunt when Hunt was putting down his first well. He had gone from field to field, from country to country, trying to make a fortune, but success had eluded him. He was in South America when he heard of the East Texas boom. He returned home and met Hunt on Henderson's main street. He told Hunt he was thinking of forming a syndicate to poor-boy a test on his mother's farm. Hunt advised him to do so, saying, "I think you'll get a big well."

Lewis drilled. The well came in a great gusher during a

heavy rainstorm. Lewis caught pneumonia and died before he could sell a barrel of oil. "Such irony," Hunt murmured when he was told. "He had hunted oil everywhere but here on his own home grounds. Now he finds it and . . ."

On a stormy night J. Malcom Crim picked up an unshaven stranger at the edge of Kilgore and drove him to the home of Lou Della Crim. Only days before Kilgore had been incorporated as a city and Malcom Crim had been elected mayor. There had been so many crimes of violence in East Texas—homicides, armed robberies and truck hijackings—that people feared for their lives. Burglars, thieves and con men worked at will. Crim hoped the unshaven stranger could change all that.

For two weeks the stranger made his way about Kilgore and the environs. His beard grew and his clothes became filthy. No one paid any attention to him; he was just another boomer. He wandered from Pistol Hill to Newton Flats, from the gambling joints in Longview to the bootlegger havens in Gladewater.

And on a bright day he surfaced. He appeared on the main street of Kilgore astride a prancing black stallion named Tony. He was booted and spurred. On his head was a white Stetson. An automatic rifle was in the saddle holster, and around his waist were slung two pearl-handled, silver-mounted six-shooters. He was one of the handsomest men Nature had ever designed—bronzed, clean-cut features, wide shoulders, slim waist. He rode the magnificent stallion like one of Jeb Stuart's cavalrymen, and he observed the street and board sidewalks with a godlike mien in which there was no pretension.

"It's Lone Wolf Gonzaullas!" a roughneck whispered—but he might as well have shouted.

"The Lone Wolf's here!" spread along the streets and into the oil field and beyond.

"He's hard, but he's fair," someone said, and oilmen who had seen him work in a dozen other Texas oil fields nodded in agreement.

The Lone Wolf was Sergeant Manuel T. Gonzaullas, Texas Ranger. He was the most feared and respected man in Texas. His tanned body bore the scars of gun battles he had fought in other boom towns and on dim cattle trails along the Mexican border. No one knew how many men had faced his guns and died, but every child in Texas had heard of his incredible strength of body and the blinding speed with which he drew his guns. He was no press agent's creation; he was "much man," as Texans described those they admired. Now, at thirty-nine, he was in the pride of his manhood.

Gonzaullas was no prude, but neither was he a civil libertarian. His mere presence in East Texas was enough to cause a minor exodus of badmen who had seen him work in other boom towns, but he paid scant attention to high-spirited roughnecks fighting in the mud streets, to the prostitutes, or to the bootleggers. He enjoyed the carnivals that pitched their tents in what open spaces that could be found, not bothering to inquire if the operators had permits. He refused to arrest an entrepreneur who, becoming aware of a great need, erected a half-dozen latrines on street corners and exacted a ten-cent payment for their use, also without a permit.

But with thieves, robbers and con men he acted as judge, jury and jailer. He had worked undercover for two weeks; now he let his presence be felt for two more. Then, on the night of March 2, he acted. With the help of other Texas Rangers he had summoned and with the aid of Kilgore's new police chief, P. K. McIntosh, whom he had helped select, he staged raids in the Kilgore vicinity.

The next morning he marched his catch down Kilgore's

main street for all to see. There were almost 300 persons in the parade. Kilgore had no jail, so he marched them to a Baptist church that had been vandalized. He already had arranged for a long, heavy chain and slender trace chains from the Crim store. His men secured the heavy chain at either end for the length of the church. The trace chains were fastened to the heavy chain like ribs on a backbone. Then the prisoners were tethered to the trace chains, which were looped around their necks and padlocked.

Police Captain Leonard Pace and two identification experts from Dallas fingerprinted the prisoners. A surprising number were wanted in Texas towns and other states.

"Lone Wolf's trotline" was an experience to be dreaded. Prisoners were fed once a day. A tin can was passed up and down the lines for a urinal. A prisoner could hardly get in a comfortable position, so crowded was the church, and the trace chain made a rough collar. Most of them accepted Lone Wolf's dictum: "I'll let you off the trotline if you get out of town in four hours." A common joke was "Lone Wolf gave him four hours and he gave three hours and fifty minutes of it back to Lone Wolf."

As fast as a batch of prisoners left town or were transferred to cities where they were wanted, a new batch was put on the trotline. At least 100 persons were so incarcerated each day until March 25, when Kilgore's new jail was ready for occupancy.

The rate of crimes of violence, thievery and all varieties of bunko games had dropped dramatically in the twenty-two days of the trotline's existence.

Gonzaullas employed another technique which seemed unfair to many. Making the rounds of pool halls and beer joints, he would approach a man he considered a suspicious character and ask what the man was doing in East Texas.

"I'm looking for a job in the oil field," the man might reply.

"Let me see your hands," Gonzaullas would say.

If the man's palms were smooth and uncallused, he went on the trotline or to jail until his background was checked out.

"Never," wrote a writer of the day, "were the horny hands of toil in more demand."

But Gonzaullas knew he was the only real "law" in the field. As the towns grew and incorporated and employed policemen, a great deal of bickering arose between the police forces and constables and sheriff's deputies. While Gonzaullas employed these men to help him when the occasion demanded, it was obvious that a number of local lawmen did not have the respect of the natives or the newcomers. Many of them failed to enforce the laws they were duty-bound to enforce. Gonzaullas concentrated on trying to rid the area of gunmen, thieves and bunko artists. He left the gamblers and prostitutes and bootleggers to be handled by local lawmen—and gambling, prostitution and bootlegging flourished.

Gonzaullas also knew he was a marked man. Every step he took was a step toward danger. Every vicious outlaw in Texas wanted his blood. It was small wonder, then, that his vigilance made him act arbitrarily. That he survived and brought a modicum of order to an almost lawless area can be credited largely to his strategy and tactics.

Gonzaullas led a raid on Newton Flats, the several-acre tent and shack community where the dance halls and gambling layouts never closed. He was not so interested in curbing the gambling, prostitution and illegal whiskey sales at Newton Flats as he was dedicated to rounding up and shipping out those he considered capable of more dangerous crimes.

His raiders took up vantage points around the community at night. At ten o'clock he gave the signal, and with their guns drawn, the raiders charged through trees and underbrush. "Stand where you are!" a leather-lunged Ranger bawled. "This is a raid!"

His shout caused a wild stampede. More than 500 persons fled through the dark, stumbling against trees, fallen limbs and vines that clawed their flesh and ripped their clothing. Of the 500, 400 made good their escape, while 100 were taken to jails where Gonzaullas sorted out the badmen from the pleasure-seekers.

He and other Rangers raided domino halls and beer joints from one end of the field to the other, keeping the pressure on. Robberies and burglaries did not cease, nor did homicides. Indeed, on occasion lawman killed lawman in quarrels that grew out of jurisdictional disputes and less savory arguments. Gladewater Police Chief W. A. Dial shot and killed a former police officer, Jeff Johnson, on the main street of that boom town—shortly after Johnson had mortally wounded Ranger Dan L. Duffie in a roaring gun battle. But without Gonzaullas and the other Rangers he would periodically call in, there would have been chaos.

Newspapers editorialized against the lawlessness, preachers preached against it and politicians inveighed against it, but little was done in the areas of gambling, prostitution and bootlegging. When a gambler, hooker or bootlegger was arrested, he or she was charged with the misdemeanor count of vagrancy. The small fine was paid gladly as a fee to continue operations, just like in big cities. Gregg County Attorney D. S. Meredith told the newspapers he was tired of the endless parade through the courtroom. He had a warning: "Hereafter, beer makers and sellers have got to face the 'Apostle Twelve' in district court on felony charges that carry a sentence of from one to five years! I'm tired of seeing vagrant fines paid and beer making and selling resumed by the offenders!"

The Longview *News-Journal* also thundered against the shady side of the boom: "There are those who apparently think gambling dens and other places of ill repute can continue to operate in Gregg County, public sentiment to the contrary, notwithstanding! This particular group is said to be brazen enough to speak disparagingly of the enforcement officers and to hurl jibes at the citizenry!"

Police Chief Sid Henderson of Longview said in an interview that "never at any time in the history of Longview have I been confronted with such a deluge of thievery. The city is over-run with an element that will steal anything they can carry or drive away." He took a slap at the Gregg County sheriff's department, saying that cooperation "could stop this run of thievery . . ."

Neither threats nor prayers, however, slowed the criminal activity, and only the presence of the Lone Wolf and his fellow Rangers kept it from getting completely out of hand.

Most of the noncriminals who flocked to East Texas in 1931 managed to make a living with their own hands and talents —something, not easily accomplished, if at all, elsewhere in the country. Wages set the level of other prices. The roughneck who made six dollars for a ten- or twelve-hour day on a drilling rig could buy a woman for fifty cents, an excellent steak and trimmings for thirty-five cents, and a pint of quality moonshine for sixty-five cents. His highest single expense was lodging, wherever he might be able to scrounge up a bed. He rested in crowded quarters on a cot still warm from the body of the previous occupant who would return to claim the cot when his work tour was over. For this the roughneck paid as much as $2 per day. Still he was able to send money home to maintain his family.

Other workers fared equally well, and small merchants fared even better. Oddly, banks were slow to exploit the boom. Their vaults were full of money from deposits and paid-up mortgages, but there were few prospective borrowers who did not want to invest in oil—and bankers considered the oil business a boom-or-bust proposition and the individuals engaged in it economically unstable. A transaction called "buying oil payments" changed their minds.

It began with an operator who had a lease but no cash to finance drilling. From a friend with cash but no lease, he borrowed $10,000 with the promise of paying the friend $30,000 out of three-quarters of the oil produced from the well. He was able to make the $30,000 payment within three weeks after the well was brought in. That became the pattern in East Texas oil financing. One dealer after another started getting $30,000 for $10,000 in a relatively short period of time. His risks were the dry hole, the blowout or the mechanical failure. There were few losses, certainly not enough to sidetrack a trend.

Some oil companies then initiated a variation on this deal. A company would simply purchase 100,000 barrels of oil to be produced for the $10,000 it would tender the operator to help finance the drilling of his well. How good the bargain was from the company's viewpoint depended on the price of oil per barrel.

Bankers became accustomed to seeing good customers borrowing money to buy oil payments. And even the staidest banker could not help but be impressed with the profits to be made in such a short time with such small risks. It was reported that the first Texas bank to lend money for the purchase of an oil payment was the First National Bank in Dallas. Others followed suit, and oil soon became as bankable a commodity as cotton, livestock or merchandise.

By the end of 1931, there were 3,607 wells in the East Texas field! More than 109,000,000 barrels of oil had been produced, and it was obvious that the 109,000,000 barrels constituted only a very small drop in a very large bucket. The oil came to the surface so readily and in such quantities that geologists and engineers were constantly revising their estimates of the giant reservoir's petroleum content.

The varying estimates confused the industry and bewildered the public. Some geologists, as a fellow scientist publicly complained, were "placing East Texas on a pedestal," while others were "belittling its possibilities."

Members of the Dallas Petroleum Geologists and the East Texas Geological Society met in Tyler on December 17, 1931, to clear the picture. Meeting Chairman Robert Whitehead of Atlantic Refining bluntly informed the scientists: "We have with us tonight several representatives of the press who will publish these proceedings. As geologists, we claim to have the facts and figures on this field, and therefore, our conclusions sponsored by the two societies should be considered significant."

Whitehead then proceeded to put the matter of the field's potential to a vote! Each of the fifty-nine geologists present wrote his estimate on a slip of paper, the estimates were totaled, and an average was struck! Some of the estimates were as low as 1,000,000,000 barrels, some as high as 3,000,-000,000. But the *average* was 2,100,000,000.

"Hell!" snorted an oilman when the story appeared in the newspapers. "That wasn't a meeting of the minds, it was a political convention!"

But 2,100,000,000 barrels! It was such a well-nigh incredible figure that the industry and the public were eager to believe it. It was accepted, for the moment. Only time would reveal that even the most wild-eyed of the geologists at the meeting had not conceived the great field's true potential.

The oil had destroyed a traditional way of life and was creating a new one. Beneath the boom-time madness a foundation for a solid future was being laid. At the end of 1931 there were 434 individual business establishments in Kilgore alone. In what a year earlier had been a hamlet of 700 souls, there were 92 restaurants, 32 supply houses, 5 welding shops, 5 boiler plants, 6 machine shops, 9 garages, 2 automobile agencies, 23 service stations, 26 lumber companies, 8 building contractors, 15 real-estate agencies, 20 dry-goods stores, 9 general merchandise stores, 11 ladies' and men's specialty stores, 12 cleaning and pressing shops, 7 hardware stores, 11 furniture stores, 35 grocery stores, 19 meat markets, 5 bakeries, 9 drugstores, 2 ice plants, 20 hotels, 25 barber shops, 8 beauty parlors.

"It's always springtime in East Texas," sang a street-corner minstrel.

But 1931 was an ugly year also, a year that saw destructive forces set loose in the great field, that saw tensions in the oil industry rise to the point of intervention by the state government.

chapter twelve

PRORATION

O il, which migrates with a change in underground pressure, will move to the nearest borehole which penetrates the oil-bearing formation. Theoretically, a single well could drain the oil from an entire field. Thus, a man with a well could produce the oil from beneath his acres, and produce the oil from beneath his neighbors' lands as well. He could do this legally under a principle called the "rule of capture," which derived from an 1889 decision of the Pennsylvania Supreme Court. The Court's judgment was based on an old English common law by which wild animals—fugitive creatures of no fixed place—belonged to "he who captures them." Oil belonged to he who drilled on lands he owned or leased and brought it to the surface . . . no matter from where it migrated.

This principle was responsible for the hasty drilling in the Texas oil fields; no one wanted someone else to get his oil. With so many wells being drilled, a field's underground pressure was gradually dissipated. Pumps would then be brought in to pull the oil from the reservoir. The first flush production from flowing wells would glut the local oil markets, and the price of crude, as it was called, would plummet. The price would rise only when demand rose because of a field's lessened supply.

But oilmen paid little heed to supply and demand. When prices dropped they produced more oil to make up the difference. Under the rule of capture they had no alternative: if an operator didn't drain his oil, his neighbor would, for strict state antitrust laws forbade him to join his neighbors in any efforts to control or restrict production.

No one knew—and few cared—if such rapid development of fields damaged the underground reservoir and inhibited production, though at one time Wallace Pratt of Humble wondered if even as much as 50 percent of the oil in a reservoir was ever brought to the surface. What happened to the oil once it was produced, however, was another matter. Crude evaporated and deteriorated in open, earthen pits where it was subject to rain and dust. Fires were frequent. Oil ran into streams and across open country. A precious, irreplaceable asset was being wasted.

A typically American hue and cry resulted, and in 1919 the Texas Legislature responded by enacting a statute prohibiting waste and requiring conservation of oil and gas. It gave the already existing Texas Railroad Commission the job of administering the new law. A few months later the commission announced some statewide rules designed to promote elementary safety. Rule Number 37 concerned well spacing, the first of its kind in the United States. It prohibited the drilling of a well within 300 feet of a completed well or of one being drilled, and within 150 feet of a property line.

Nothing was said about the rule of capture, however, so unrestrained drilling and production continued. And the commission did not enforce diligently the rules it had set up. Even basic Rule 37 could be circumvented by an operator or by a company who obtained an "exception" from the commission . . . and almost every operator and company did, generally using the rule of capture as the reason.

Most oilmen—those from the major companies in particular—despised government control on any level unless such "interference," as they called it, worked to their advantage. In this they differed little from other industrialists. The commission, for example, was empowered to prevent the waste of gas produced simultaneously with the oil. Most oilmen felt that in the course of producing oil they were entitled to blow any amount of gas into the air. There was no commercial market for gas in Texas until 1926, and that was still a small one. Oilmen generally believed that gas was not being wasted, even when it was blown, if it had served to lift the oil from the reservoir. Since the commission was not inclined to punish violators, most oilmen continued to do as they pleased.

But in 1923 the entire industry came under attack from several quarters. After a long investigation, a U.S. Senate committee reported that certain large companies had conspired to rob both the small producer and the public. The Teapot Dome scandal was aired that same year, and fear of an oil shortage focused attention on the industry's wasteful practices. There were calls for federal regulation, one coming from one of the country's leading oilmen, Henry L. Doherty, chief executive officer of Empire Gas and Fuel Company.

Doherty was undoubtedly one of the best-informed professional oilmen in the land. His intellectual curiosity had led him to study the precious little material available on the behavior of oil and gas reservoirs. He had followed closely the work being done at the Bureau of Mines Petroleum Experimental Station at Bartlesville, Oklahoma, and had financed research aimed at determining the relationship between gas and oil in a reservoir. He had concluded that it was vital to utilize gas to drive oil through reservoir rocks.

In his arguments for federal regulation, Doherty maintained that proper conservation of gas would make possible the recovery without pumping of virtually every barrel of oil from a pool. He decried excessive drilling, and called for storage of the oil in its natural habitat, to be produced as needed, in order to prevent waste and to maintain a stable price structure.

Most oilmen denounced his ideas. In 1924, however, the Federal Oil Conservation Board was established, and President Calvin Coolidge invited oil leaders to participate in a study of industry problems. The American Petroleum Institute, official spokesman for the industry, adopted a resolution offering full cooperation, and appointed a committee to study the problems of the industry.

The committee's report contained no surprises for oilmen. Waste, said the committee, was negligible. There was no imminent danger of exhausting U.S. petroleum reserves. The availability of future supplies depended on adequate incentives such as "security in ownership of lands and the right to lease; conditions permitting exercise of initiative, liberty of action, the play of competition and the free operation of the law of Supply and Demand; and prices that will provide a return to producers, refiners, and distributors commensurate to the risks involved and the capital invested."

Every member of the committee and every oilman in the country knew that the key statement in the report was a lie, and so did every interested citizen and politician, including the President. There *was* great waste and, under existing drilling practices, a grave danger of exhausting reserves. The remainder of the report had little relation to the problems, which were real, but appeared to be the outline for a freewheeling oil industry manifesto.

What the major companies actually wanted in Texas was repeal of the state's strict antitrust laws. By pooling interests in a field, their spokesmen suggested, exploitation of the

field could be accomplished in an orderly manner—and neither state nor federal regulation would be necessary. The
independents saw the suggestion as an effort by the majors
to freeze them out. They *had* to keep drilling and producing
in a field, they maintained, to stay in business. Otherwise
they would be forced to sell out their interests at the majors'
price.

Doherty sympathized with both sides, but he had little
hope that oilmen would voluntarily respond to any plan
aimed at conservation. Federal regulation, however, would
tend to be fair and just while saving a precious national asset,
he argued.

Though Doherty's ideas were publicly denounced, they
took hold to some extent in oil company board rooms. That
better production methods would help create a more stable
price structure was obvious. And the old-line oilmen were
learning from a new breed—especially petroleum engineers
—that because of existing production methods, great quantities of oil were being left in the reservoirs, and could not be
recovered thereafter.

While company executives were being educated by their
petroleum engineers, their geologists and geophysicists
were finding new oil fields. By 1926 numerous new fields in
the United States began to add to the world's oil supply. And
oil was flowing from the Maracaibo Basin of Venezuela, from
Colombia and from new fields in the old oil province of
Sumatra. Stocks of crude and refined products in the United
States soared. Prices began to decline. Some oil executives
who before had considered the words "government regulation" industrial blasphemy began looking for governmental
help.

Foremost among these in Texas was W. S. Farish, president of Humble Oil and Refining Company. Farish had long
and stoutly contended that the industry could handle its
affairs without government interference. Very early he had
condemned the efforts of Doherty and others to eliminate

wasteful practices. But once Farish made the about-face, he began campaigning for gas conservation and unitized field production—under state control—with all the brilliance and energy that he had displayed in building Humble into a petroleum giant. Two other leading oilmen, J. Edgar Pew of Sun and E. W. Marland of Marland Oil Company, joined forces with him.

Under Farish's guidance, other Humble executives made educational speeches, wrote tracts and pushed hard for conservation legislation. They were rewarded with little success. But in the autumn of 1927, Humble took the lead in a move to prorate production of a large oil field: the Yates pcol in West Texas. It was the first time such an attempt had been made in the state.

Humble had the only pipeline in the vicinity of the pool. At that time, the field could produce 192,000 barrels of oil per day. Humble offered to extend its line to the pool and buy 30,000 barrels daily, the amount it estimated it could market, if the operators would agree to ratable sharing of the outlet. Farish bluntly pointed out that without such an agreement much of the oil could not be marketed and that it would be expensive to store. And uncontrolled production, he added, would result in waste—with consequent loss to the public—and in financial disaster to the producers.

There were arguments, charges and countercharges before the Yates operators agreed upon a plan of operations. The plan was far from satisfactory for all concerned, but by June 1928, 150 wells on 14,000 acres, capable of producing 1,800,000 barrels a day, were being held to a production of 52,500 barrels. Gas was conserved and reservoir pressure remained strong.

Farish kept up his campaign, urging the Railroad Commission to introduce proration in other fields as the operators were doing voluntarily at Yates. The commission responded by issuing orders limiting production in Winkler County to 150,000 barrels daily, prorated according to a formula giving equal weight to acreage and well potential. Later it assumed the administration of the Yates field, where it established 100 acres as the proration unit, allowables to be figured one-fourth on an acreage basis and three-fourths on the basis of well potential.

This was the first step in state regulation of production, albeit a timid one. Other Texas fields were being drilled and produced as recklessly as before. Farish continued his campaign. It will be recalled from Chapter Four that in the autumn of 1929, Humble, Pure, Shell, Gulf and the Texas Company had agreed to operate the Van field in Van Zandt County as a unit. It was this field that briefly stimulated interest in East Texas and prompted Dad Joiner to renew his efforts to drill the Daisy Bradford 3 in Rusk County.

Farish's efforts were responsible to some degree for the organization of the Central Proration Committee, a group of oilmen from all producing segments of the industry who began working for voluntary proration as practiced at the Yates field.

But Farish and Humble lost friends and supporters in January 1930, when Humble instituted a drastic cut in the price of crude, saying the cut was necessary because of low gasoline prices resulting from excessive stocks. The independents who had cooperated with Farish in the proration movement had done so with the tacit understanding that crude prices would be maintained. Now they felt betrayed.

Farish offered arguments which appeared valid to some, but at both state and federal levels groups of angry independents pushed for legislation aimed primarily at Humble and Standard Oil Company of New Jersey, Humble's majority stockholder. Independents blamed much of the overproduction woes on imports, and Jersey Standard was a large importer of foreign oil. A bill calling for a dollar-a-barrel tariff on foreign crude was introduced in the U. S. Congress. It lost by a small margin.

Autos jammed the road leading to the Daisy Bradford 3 when Ed Laster was trying to bring in the well.

Dad Joiner and Doc Lloyd congratulate each other in front of the completed Daisy Bradford 3. The two white-shirted men on the right are H. L. Hunt (with straw hat and cigar) and Ed C. Laster. The crew, from left to right, are J. Sistrunk, James Hunt, D. Hughes, Glenn Pool, J. P. Maxwell, W. A. Kirkland. Dennis May, a faithful semi-regular crewman, was not present for the photograph.

Doc Lloyd, Daisy Bradford and Dad Joiner, in front of the soda-water stand Miss Daisy's nephew built to slake the thirsts of the thousands who witnessed the birth of the Black Giant.

Walter and Leota Tucker. They never gave up on Dad Joiner.

Ed C. Laster. He drilled the Daisy
Bradford 3.

E. A. Wendlandt. His recommendations
kept Humble in East Texas.

H. L. Hunt as he looked during the boom.

J. Malcolm Crim. His belief in a fortuneteller made him wealthy. The corncob pipe was his trademark.

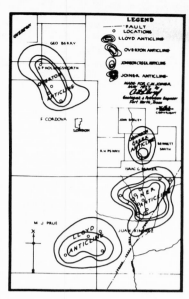

Doc Lloyd's map which accompanied his report to Dad Joiner. The fault line and anticlines he drew did not exist. Map and report were Joiner's sales tools.

Barney Skipper. The Longview Prophet, responsible for the third producing well in the field, the Lathrop 1.

W. W. Zingery, the mapmaker who provided the independents with ownership maps of Rusk County, and who helped initiate the receivership suit
against Dad Joiner.

Lou Della Crim, the gra-
cious lady on whose farm
the second well in the
field was drilled.

Wallace Pratt, Humble's brilliant geologist.

W. S. Parish, Humble president and a leading fighter for proration.

Walter Lechner. He put the Lou Della Crim 1 deal together for Skipper

Barney Skipper, driller Andy Anderson and F. K. Lathrop in front of the Lathrop 1, the third producing well in the field.

Small part of the vast crowd that witnessed the bringing in of the Lathrop 1. More than 18,000 were on hand when the well came in.

D. H. (Dry Hole) Byrd on the derrick floor of one of his wells, the Elder 1, in Gregg County, on March 5, 1931.

Doc Lloyd stands in the middle of a happy group at the Tex Lloyd 1, a well promoted by Tex Lloyd, the young man in boots and white trousers at the left. The smiling youngster second from the far right is Doug Lloyd. He roughnecked on the well. He revered his father and fought to have him regarded as a geologist rather than as a promoter.

Kilgore's population jumped from 700 to 10,000 within days after the completion of the Lou Della Crim 1.

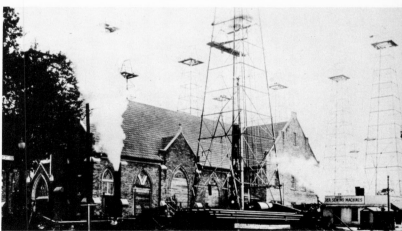

Even churchyards were drilled at the boom's height. This one was in Kilgore.

A forest of derricks sprung up
in Kilgore during the boom.

The "world's richest half block"—twenty-four wells were drilled on this half block along Kilgore's Commerce Street. There is no oil beneath the half block now, and only one derrick stands, as a commemorative marker, the one in the lower right-hand corner. A parking lot is on the half block.

Ranger Sergeant M. T. (Lone Wolf) Gonzaullas astride his horse Tony and wearing his automatics, not his usual silver-mounted six-shooters.

General Jacob Wolters and Ranger Captain Tom Hickman reading Governor Ross Sterling's martial law order for the East Texas field on the steps of Kilgore's city hall.

Texas Rangers called in to help troops close the field. From left to right, Captain Tom Hickman, Sergeant M. T. (Lone Wolf) Gonzaullas, Ranger W. E. Young, Ranger Bob Goss, Ranger S. Stanley and Captain Albert Mace. Gonzaullas had kept an unsteady lid on crime almost alone until martial law was declared.

E. O. Buck, the engineer who pointed the way to the field's salvation.

Carl Estes, fire-eating editor and a leader of the
anti-prorationists.

Tom Kelliher shut off "hot oil" flow.

J. Howard Marshall, a champion.

Ernest O. Thompson, the "Redhead from Amarillo" as he grayed and grew in stature.

F. W. (Big Fish) Fischer. He took the "hot oil" runners' case to the Supreme Court—and won.

Judge R. T. Brown, the "Sage of East Texas."

The independents had better luck in Texas, however. On March 18, 1930, the Texas Legislature passed what was called the Common Purchaser Act. It required any purchaser of oil in Texas affiliated in any way with a common-carrier pipeline to buy oil ratably without discrimination between fields or producers—including the purchasing company's own production.

Even though the new legislation had been designed to punish Humble, Farish was not at all unhappy with it. He reasoned that the act would help bring proration to the state because oil production would necessarily have to be limited to the purchasing companies' needs. Certainly under this law the Railroad Commission could, if it desired, restrict production to market demand.

Then, on August 14, 1930—while Dad Joiner was still drilling—the commission issued its first statewide proration order. It limited Texas production to 750,000 barrels daily—an amount it estimated would supply a *reasonable market demand*.

Immediately a number of companies sought temporary injunctions that would allow them to produce without restrictions. One of these was the Danciger Oil and Refining Company of North Texas. The company argued that no law could prevent a company from producing as much oil as it had a market for; that the commission's order was not concerned with physical waste, but with price fixing and economic waste, which were expressly excluded under the meager conservation laws of the state.

The Danciger temporary injunction was granted. Proration broke down and the Texas Railroad Commission's newly won prestige evaporated. Legal confusion reigned as overproduction increased, prices fell and the Great Depression got a firmer hold on the nation's economy.

Against this background, Dad Joiner brought in the Daisy Bradford 3 on October 5, 1930. This was less than two months after the commission's statewide proration order and less than a week after Danciger obtained a temporary injunction against it. On the average, oil was selling for $1.10 a barrel, a sharp decline from the $2.29-a-barrel price in the pre-overproduction year of 1926.

Then Ed Bateman brought in the great Lou Della Crim 1 on December 28, and on January 26, 1931, the Lathrop 1 came in. A month later, with wells roaring in around the clock in the great new field, oil was selling for $.50 a barrel and prorationists were screaming for order.

But even with the turmoil of the boom and the proration battle shaping up, East Texas was still able to be shocked by the news that Dad Joiner had gone to court in Henderson to sue his now-famous driller, Ed Laster, and the Mid-Kansas Oil and Gas Company, alleging that they had conspired to defraud him.

If Joiner had a right to be angry with Laster, he didn't have much of a lawsuit. He contended that Laster had been duty-bound to keep details of the well-drilling secret from everyone but Joiner; that Laster had given such details to Mid-Kansas in return for an undivided quarter interest in the 1,100 acres the company had leased on the basis of the details; that Laster had told Mid-Kansas about taking the core from the Daisy Bradford 3 before he had told Joiner; that Mid-Kansas had leased the 1,100 acres for $1 an acre before Joiner could do so—and the leases had risen in value to $3,000 an acre or a total of $3,300,000.

Joiner wanted the leases. He was willing, he said, to pay Mid-Kansas whatever the company had paid for them.

Laster answered that he had not been under contract to Joiner and was thus free to acquire leases at any time he chose. He said he had not told Mid-Kansas or anyone about taking the core before he told Joiner. But, he said, some days after the core was taken, Joiner himself had showed samples of it on the streets of Tyler. And Joiner had made no effort to buy the disputed leases or any other leases. "There were thousands of acres he could have leased if he had so desired," Laster said. "But he always said he was broke. He was trying to sell leases."

Laster admitted making the deal with Mid-Kansas, but said that he had not received an undivided quarter interest in the 1,100 acres. After some wrangling, with him threatening to sue, Mid-Kansas had given him $10,000 and a small lease not included in the 1,100 acres.

Laster also claimed that Joiner had not delivered him leases he had earned while drilling the Daisy Bradford 3. He had worked for $6 a day in cash plus $4 a day in leases. Dad, he said, still owed him leases totaling 370 acres and worth $149,500.

Joiner and Laster reached some kind of agreement that was not made public. In an amended petition, Joiner did not list Laster among the defendants. Mid-Kansas then asked that the case be transferred to federal court since the plaintiff and defendant were domiciled in different states.

Judge R. T. Brown granted the motion. The case was transferred to federal court in Tyler, where it was promptly dismissed. Mid-Kansas drilled the acreage and became one of the field's large producers.

The issue between Joiner and Laster was essentially a moral one. Joiner felt that he had been betrayed, but he apparently couldn't prove it. And had he been able to prove it, it very likely would not have constituted the basis for a victorious lawsuit. Laster obviously gave Mid-Kansas vital information in time for a company with resources to acquire cheaply leases which overnight had become desirable ones. Oilmen would not fault Mid-Kansas for what could be considered a brilliant business coup. If Laster could be questioned, it would be on the basis of the degree of loyalty he had owed Joiner.

chapter thirteen

THE BATTLE IS JOINED

Three men sat on the Texas Railroad Commission, elected by the people to staggered six-year terms. The commission was a haven for retired or defeated politicians whose names were still familiar enough to the citizenry for them to be considered for a job in the backwaters of politics. At the beginning of the wild boom in East Texas the commissioners were Lon A. Smith, C. V. Terrell and Pat Neff, a former governor. Neff was commission chairman.

The commission's orders to the oil industry were to be carried out by the Oil and Gas Division, now the commission's most important arm. This department had representatives in the oil-producing areas of the state, normally an umpire and supervisors whose duties were to enforce the commission's regulatory orders. Ineffective at best, the commission and its Oil and Gas Division were not able to cope with the challenges presented by the chaotic conditions in East Texas, and generally did not seem to care.

The governor of the state was Ross Sterling, a founder and former president and board chairman of Humble Oil and Refining Company. He had been elected in 1930 and had begun his two-year term of office on January 10, 1931. Oil and proration had not been an issue in the campaign. His chief opponent had been Miriam A. (Ma) Ferguson who had sought office as a vindication of her husband, James A. Ferguson, an ex-governor of Texas who had been impeached during a resounding state scandal. So Sterling had run against "Fergusonism" and as a level-headed businessman capable of leading the state during depression times.

But hardly had he entered the governor's mansion in Austin when oil and proration became the dominant issue of the day. His background with Humble had been an asset in his campaign: was he not a builder, the owner of the Houston *Post-Dispatch*—and had he not been the chief executive officer of the state's largest oil-producing and oil-purchasing company? He had left Humble in 1925, but because Humble's 1930 price cuts had infuriated most of the independents and their business and political allies, Sterling was quickly labeled Humble's creature, a tool of big business.

It is doubtful that Sterling deserved the appellation. He had left Humble amidst rumors that he was not on the best of terms with other Humble officials. And his actions during the first months of the boom were so confusing that it was not possible to determine his attitude on proration. In this regard, he was as ineffective as the Railroad Commission members, who made no attempt to regulate production in East Texas until April! By that time more than 160,000 barrels of oil daily were being produced and a new well was being spudded in every hour.

Rule 37, which prohibited the drilling of a well within 300 feet of a completed well or one being drilled, and within 150 feet of a property line, was ignored by most of the operators, large and small. Those who wanted to maintain at least an air of legality obtained exceptions to Rule 37 from the commission in wholesale batches. The commission had no alternative under its own rules—and it would not change the rules. In one instance it allowed twelve wells to be drilled on a 60-by-150-foot lot in Kilgore!

It must be remembered that when the Daisy Bradford 3 was brought in, the poverty-stricken natives of East Texas knew little or nothing of the oil industry's problems. To them the discovery well promised blessed relief from the economic misery they had shared for a decade. In the sudden flow of oil they saw an end to their bondage to capricious crops and weather. The first public statement about restricting the field's production, though it came from an anti-proration group, was a dagger in their entrails.

The oil publications had touched lightly on the subject, saying that a sizable pool in the Joiner area might damage the already weakened oil price structure. But the natives didn't read oil publications. However, almost every adult in the oil province did hear in some fashion about a resolution passed on January 2, 1931, by the Marion County Chamber of Commerce. Marion County had no direct stake in the field—it was well outside the apparent oil zone—but its businessmen told the Central Proration Committee to keep its "hands off" of East Texas "in view of the fact that this section is just entering the possibilities of the largest oil field in the state." The Central Proration Committee, it will be recalled, was the group attempting to limit production in the state by voluntary means.

The war cry from Marion County was picked up by the keen ears of Carl Estes, the Tyler editor who had attacked the "big boys" and "slick lawyers" whom he had seen moving in to steal Dad Joiner's oil even before Joiner had completed the Daisy Bradford 3. Estes was a populist, and he knew his East Texas readers. He was now forty-five, a man of medium size who at times seemed almost shriveled by the near-constant pain he suffered from an ulcerated stomach. His ailment forced him to walk with crutches occasionally.

Estes detested his ailment because he was a hearty man, and he loved a good fight, as he once remarked, "more than a hog loves slop." Now he saw a good one brewing, and he quickly organized a citizens' committee whose avowed purpose was to study proration.

But on January 11, a full-page advertisement—or something laid out as an advertisement—appeared in both the Tyler *Morning Telegraph* and the Tyler *Courier-Times*, the afternoon newspaper. But it was a scathing denunciation of the "monopolistic companies" who were advocating proration "as a means of holding up the price of oil." The "advertisement" demanded, "Has the time arrived when the 'big boys' of the oil industry can tell the people of Texas who may produce oil, when they may produce oil and what oil prices should be? We say NO."

The battle was joined when copies of an article by Robert R. Penn in the *Oil and Gas Journal* were widely circulated. Penn was an independent oilman and chairman of the volunteer Central Proration Committee. In the article Penn eschewed belligerence while outlining the merits of proration. But he bluntly warned that "if the operators enter into a wild and reckless drilling campaign, the problem [overproduction] will be greatly exaggerated, and there will be many wells that will never pay for themselves even though the area might develop into a rich one."

Humble then took up a cudgel. On January 15 the company announced that it was prepared to enter the field as a purchaser. It already was a producer by virtue of buying the Lou Della Crim 1 and Ed Bateman's interest in the surrounding acreage. Humble said it already had a pipeline from the Van field to the Louisiana border which it planned to extend to the East Texas field to handle its own oil. It would then buy as much oil from others' leases with which its pipeline could be connected as it would run from its own leases—on the condition that orderly, ratable production be established. It would not buy from others as long as production was unrestrained.

This was a hard blow to the independents who needed outlets for their oil. At a five-hour meeting in Tyler, the feelings of most of the independents and natives were summed up by District Judge Charles Brachfield. "When other fields were turning out millions of barrels of oil, nobody thought of proration," Judge Brachfield thundered. "It certainly ought not to start here until our millions of barrels are developed. But now, before we have shipped our first hundred thousand barrels, they want to stop us with proration!"

Meetings followed meetings. Associations were formed, merged, disbanded and formed again. Heavy oil imports were blamed for the glutted market. So was a world-wide monopoly. It was true that Gulf, Shell and other majors had great oil fields in foreign lands, and brought much of that production into the United States. Humble's chief stockholder, Standard Oil Company of New Jersey, operated around the globe, and could bring oil from Venezuela to its East Coast refineries as cheaply as it could transport oil from Humble's wells. But it was also true that unrestrained production from Texas alone could almost satisfy the U.S. market—and unrestrained production meant the loss of millions of barrels of oil that could not be recovered from damaged reservoirs.

But oil imports were not to be curtailed or subjected to high tariffs; the majors had sufficient political clout in Washington, D. C., to prevent that. So the independents' chief problem boiled down to this: Did they want to prorate the field and sell their oil to willing buyers for a dollar a barrel, or did they want to go their merry way and sell their oil— when they could find a buyer—for twenty cents a barrel, or less?

Some independents began swinging to the side of the prorationists, but the great majority held fast. Both sides maintained a steady propaganda barrage. The Texas Railroad Commission announced it would hold a hearing on East Texas production on April 1, 1931. Prorationists demanded an earlier hearing. Anti-proration groups from every area of the field met in Longview on February 5 and formed the East Texas Lease, Royalty and Producers Association. Judge Brachfield was elected president. W. B. Hamilton, a Wichita Falls independent operator, was elected vice-president. And Carl Estes, the firebrand, was named secretary–treasurer.

In a resolution the group urged the commission to "refrain from entering proration orders of any character effective in the newly discovered Gregg and Rusk Counties pool or pools until such time as the area of said field is fully defined and production therefrom is sufficiently advanced." It stressed that in their early development, other fields had been allowed flush production "and we see no reason why the Rusk and Gregg Counties pools should be made an exception to the rule."

On the same day in the same town another group, the Proration Rules Committee of the Gregg and Rusk Counties Proration Advisory Committee, met and delivered a statement in opposition. This group urged the commission to call a hearing "as soon as possible" in order to prorate East Texas production in relation to market demand. In this group was H. L. Hunt, who with the purchase of Dad Joiner's five thousand acres had become one of the largest leaseholders in the field. Hunt was charged with having "sold out to the majors" by fellow independents, other leaseholders, and merchants and farmers. But Hunt had fought for proration in Arkansas and Louisiana before coming to East Texas, and now he had more to protect than ever before.

The *Oil and Gas Journal*, whose reporters covered both meetings, warned in its subsequent issue: ". . . if proration is not put in force in these new areas, lack of pipeline outlets and market demand will in itself curtail production and drilling, and distress oil prices will prevail, having already entered the areas and causing the crude to be marketed at prices far below what it would ordinarily bring in better times."

Estes kept hammering away at proration in his newspapers, calling it premature and blasting its proponents. Where an *Oil and Gas Journal* poll found a "majority" of operators favoring some sort of proration, a poll conducted by the newspapers found the exact opposite.

So it was that on February 8, the "country editor"—as he styled himself—was called on in Tyler by three prominent representatives of the major companies: Walter Teagle, president of the Standard Oil Company of New Jersey; W. S. Farish, Humble's president; and John Suman, Humble's brilliant petroleum engineer who had educated Farish and other Humble leaders away from their wasteful practices of earlier days.

What was said in the meeting was not disclosed ; perhaps Suman attempted to educate Estes. In any event, the four men agreed to formation of a twenty-man committee charged with working out a solution to the East Texas crisis. The committee was formed. It was composed of five independent oil operators, five land and royalty owners, five major oil company representatives and five crude purchasers not affiliated with the majors.

Before the committee could elect officers, District Judge C. A. Wheeler rendered a decision in the Danciger case in Austin. Danciger Oil and Refining Company, it will be recalled, had obtained a temporary injunction against enforcement of the Railroad Commission's statewide proration order of August 14, 1930. The injunction had caused the breakdown of proration and was a blow to the commission's authority and prestige. Now, on February 13, 1931, Judge Wheeler denied the company a permanent injunction, thus affirming the authority of the commission to prorate production under the police powers of the state.

Danciger immediately appealed the decision, leaving the issue in doubt since the case would not be settled for some time. Humble announced it would continue building facilities to handle oil from its leases, but would not buy oil from others. Other companies, however, were more willing to purchase, particularly Sinclair, which had bought the first oil from the field through H. L. Hunt's pipeline from the Daisy Bradford 3. But the independents needed Humble and every other purchaser they could find, so great was the field's actual and potential production.

Now Governor Sterling began receiving calls which urged him to call a special session of the Texas Legislature to enact a market demand law. Sterling did not budge. Then in March he attended a meeting called by Governor W. H. Murray of Oklahoma to consider oil-industry problems. Representatives of the governors of Kansas and New Mexico participated. A report was issued that recommended proration based on market demand and the application of that principle to the East Texas field.

While scores of drilling rigs operated night and day, and wells roared in, each producing thousands of barrels of oil, the Texas Railroad Commission announced that a special hearing on East Texas would be held in Austin on March 24, not April 1. The East Texas Lease, Royalty and Producers Association met and expressed the group's willingness to accept proration if oil imports were reduced to 16,000,000 barrels annually, an amount recommended in a bill before the U.S. Congress. "The great oil combines should not be permitted to crush the small independents, the men who made possible the great East Texas field!" shouted Estes the newspaperman.

Attorneys were hired—including Dan Moody, a former governor—and plans were made for a protest march on Austin on the day of the commission hearing. On the appointed day the marchers arrived in Austin in thirteen Pullman cars. Each side presented its arguments, Dan Moody speaking for the anti's, Robert Penn of the Central Proration Committee speaking for the pro's. Penn, an independent oilman, had long been called a stooge of the majors. At the hearing, Moody asked him to tell truthfully what his connections with the majors really were.

Penn went after Moody, his fists clenched, his face flushed with anger. Carl Estes proved he couldn't stay out of any fight; he struggled to his feet, waved a crutch and shouted at Penn: "Come on, you son of a bitch, and I'll knock your brains out!" Others jumped up to calm the belligerents. Penn denied he had any personal interest in any major company, insisting that he was working for the good of the entire industry.

It was not until April 4 that the commission was able to set up an order for the field, to become effective May 1. It set the initial total daily allowable at 50,000 barrels, changed that to 90,000 and when May 1 arrived had finally decided on 160,000 barrels. The final decision divided the field into parcels of twenty acres or fractions thereof. The allowable of each parcel was to be determined by the relation that its potential bore to the potential of the field, except that every parcel or fraction of a parcel was to be allowed to produce at least 100 barrels daily. But the commission had said nothing about curtailing drilling, so more wells were drilled to increase a parcel's total production.

Few operators obeyed the proration order in any fashion. Some brought injunction suits against the commission and, behaving as if the mere filing of the suit provided protection, continued to produce to capacity. The pitifully few inspectors sent into the field by the commission were helpless.

By the end of May the majors were paying only fifteen cents a barrel for oil, others were paying as little as six cents, and some sales were being made for only two cents a barrel! Still the drilling orgy continued. During the week of June 7, Humble led all other operators in obtaining drilling permits. And by now there were 1,000 completed wells in the field, and 500,000 barrels of oil daily were being produced!

And the oil was being marketed! Majors were buying it because it was cheaper than oil in other fields. Wells in other fields, more costly to operate, were being shut down. Brokers were buying oil and shipping it to distant markets. And some shrewd men, seeing that the oil was not always moved freely to the refineries of the major and large independent companies, moved into the field and built refineries! Six were in operation or nearing that point in June. The number rose to more than ninety-five, then stabilized for a period at seventy-six. It was the greatest influx of refineries ever concentrated in a single area! Indeed, there were more refineries in the East Texas field than in the rest of the state of Texas and probably the United States.

In addition, the country's most extensive railroad facility for moving oil was mushrooming in and around the field. Seventeen trunk pipelines and thirty-five gathering systems were in operation or near the final stages of construction. Even Humble, using the commission's May 1 proration order as its reason—or excuse—had tied in a number of independents' wells to its gathering system—something it had announced it would not do as long as production was unrestrained.

While the first refineries provided a market for the independents' crude, they also brought great profits to their owners. Because East Texas crude was of such high quality and contained such a large amount of "light ends," gasoline could be produced with the most rudimentary refining equipment. A small refinery could be erected at costs ranging from $10,000 to $25,000. Crude cost fifteen cents a barrel. Gathering charges, refining and marketing expenses totaled thirty cents. Sixteen gallons of gasoline could be refined from the barrel of crude—and the gasoline sold for four cents a gallon, thereby grossing sixty-four cents for the gasoline refined from a barrel of oil. With a net profit of nineteen cents per barrel of crude processed, a refiner processing 30,000 barrels a month could make a net profit of $5,700 on a $10,000 to $25,000 investment!

A few of the first refineries were substantial ones, but the great majority were called "teapots," the owners of which were interested only in "skimming out" the gasoline from the crude. The refineries also brought another industry into being. Great fleets of trucks roared out of East Texas around the clock, carrying gasoline to cities, towns and whistle stops within a 300-mile radius. The gasoline was low in octane, but it would run a car—and it was cheap. It sold at service stations for as little as eleven cents a gallon, much less than the majors and large independents were charging for their regular grade. "Eastex" gasoline became a byword in the state as small service stations sprung up to sell the product and long-closed stations were opened.

Despite the crude price decline and the Depression, the majors and large independents had cut the prices of gasoline and refined products at their service stations only a fraction. Now the cheap gasoline from the East Texas refineries began cutting into their sales of regular-grade gasoline so deeply that the majors and large independents were forced to begin marketing their own Eastex gasoline. Stations selling the original retaliated by offering a free chicken dinner, a dozen eggs or a crate of tomatoes with a fill-up.

All of this business activity added up to a good argument for the anti-prorationists. In a country in which the economy had ground to a standstill, thousands of men were making a living in the East Texas field. Thousands more were making a living servicing them—farmers, grocers, druggists and the like. In other cities, machinists were making oil-field equipment because of East Texas; steel mills were turning out pipe and material for workers to use in fabricating drilling rigs, barrels, and trucks. Railroads were busy.

This may have been strong in Governor Sterling's mind during the tumultuous months when he ignored an avalanche of telegrams from prorationists urging him to call a special legislative session for enactment of a market demand law. He also knew that Texans had been victimized by trusts and monopolies in the past, and that proration was being equated with the "big boys." And he was aware that most Texas businessmen—including himself—had a deep-seated resentment of any type of government regulation.

In any event, Sterling waited. He waited while the commission tried to enforce a second proration plan containing the provisions of a plan offered by an independent oilman, Tom Cranfill. He waited while this plan failed, and oil dropped to ten cents a barrel in other fields around the state.

Finally, on July 14, 1931, the Texas Legislature was convened in special session in Austin. A bill was quickly introduced that specifically authorized the restriction of production to reasonable market demand. Its proponents argued that such a restriction was vital to the prevention of physical waste. The opposition saw the bill as a scheme to promote price fixing by the "big boys." An investigation of the oil industry was demanded. The investigation—if it could be called that—lasted almost a month. It was marked by name-calling on both sides as witness after witness was called to testify and wound up being vilified.

While this side show was in progress, on July 28 a three-judge federal court handed down a shattering decision. The case was that of *Alfred MacMillan* et al v. *the Railroad Commission.* The court opinion, written by Judge Joseph C. Hutcheson, held that the April 4, 1931, commission order prorating the East Texas field was invalid—that it had no reasonable relation to physical waste, but was based on market demand, which related to economic waste, and thus was in violation of the state's basic conservation law. The opinion discounted as speculation the testimony of geologists and petroleum engineers that dissipation of reservoir energy resulted in physical waste. Proration, said the opinion, was only a device to solve the problem of glut.

The legislature, however, had the authority to enact a market demand law—one related to economic waste—but Sterling suddenly announced that he would veto any bill that authorized restriction of production to market demand. Since the special session had been called by Sterling to consider just such legislation, the governor's announcement, while bewildering the prorationists, gave strength to their opposition. The anti's pressed their case and the legislators began falling into line.

Then, on August 4, Oklahoma Governor William (Alfalfa Bill) Murray ordered the Oklahoma City and Greater Seminole fields closed in by state troops. Murray flatly declared that troops would remain on guard and that the wells would stay shut down until oil again sold for a dollar a barrel. The Oklahoma City and Seminole fields were large ones with flush production, but East Texas was ten times larger than either and thousands of times richer.

As if unaware of Murray's action, the Texas Legislature obliged Sterling by enacting a statute specifically prohibiting any restriction of production to market demand, although it did prohibit almost every kind of physical waste imaginable. This was on August 12.

On August 13 it became known publicly that thirty-seven East Texas operators had sent Governor Murray a congratulatory telegram for shutting down the Oklahoma City and Seminole fields with state militiamen. The telegram's senders regretted "that the same fine character of leadership and courage has not been shown in the State of Texas. We feel that had such leadership been exercised here, conditions existing today would have been averted."

On August 14 the East Texas Chamber of Commerce sent Sterling a resolution asking him to declare martial law in "the oil producing territory of Gregg, Smith, Rusk and Upshur Counties." The resolution was accompanied by a petition with 1,200 signatures.

On August 15 it was announced that the East Texas field was producing one million barrels of oil a day!

And the next day Governor Sterling declared martial law in East Texas and sent in the National Guard to enforce it.

chapter fourteen

MARTIAL LAW
AND HOT OIL

As a founder and early leader of Humble, Ross Sterling had been considered a brilliant oilman. As a governor dealing with the most vexing problem in the industry's history—one that was racking his state—he appeared to have forgotten the lessons of his youth. For months he had vacillated while chaos increased in East Texas. He had called a special legislative session to consider effective legislation, then had destroyed that promise with a statement which was a repudiation of the common sense with which for years he had been credited. With the declaration of martial law in East Texas, he had taken a bold, decisive step. Now he promptly did everything in his power, it seemed, to render it an impulsive gesture.

Whereas Murray of Oklahoma had sent in his troops to obtain "dollar oil," giving no apologies or excuses, Sterling, in his proclamation, declared: "There exists an organized and entrenched group of crude petroleum oil and natural gas producers in said East Texas oil field, covering areas within Upshur, Rusk, Gregg and Smith Counties, who are in a state of insurrection against the conservation laws of the state relating to the prevention of waste of crude petroleum oil and natural gas and are in open rebellion against the efforts of the constituted civil authorities in this state to enforce such laws."

"Insurrection" and "open rebellion" were, at the least, badly chosen words. And in remarks to newspaper reporters, Sterling spoke of a possible "outbreak" in East Texas without saying who was threatening one.

Further, he selected as troop commander Brigadier General Jacob F. Wolters. Jake Wolters was very experienced in establishing and maintaining martial law, but he was also general counsel for the Texas Company (Texaco). One of his top aides, Colonel Walter Pyron, was a Gulf official. If the anti-prorationists had needed more propaganda ammunition to shoot at the "big boys," Sterling certainly had supplied it.

Wolters was a balding, burly man of sixty. He was a good lawyer, but a politician whose success had never matched his ambition; he had won election to county attorney and state representative, but had lost a race for the U.S. Senate. He loved both the law and the military. He had been a first lieutenant in the Texas cavalry unit he had helped organize during the Spanish-American War, but he didn't get overseas. In June 1919, however, he and his troops had stopped violent clashes between whites and blacks in Longview, and had resolved the issues between them. Later that year he commanded troops in the Corpus Christi area after a violent hurricane caused widespread destruction.

The following year martial law had been declared in Galveston, where a shipping strike had caused outbreaks of violence, and Wolters was in charge of the troops that brought order. In 1922 he had commanded troops in Mexia when local officers could not enforce the laws in that wild oil boom town. And he led troops into Borger, another oil boom town, in 1926. In neither Mexia nor Borger were the fields shut down.

In the early morning of August 17, 1931, Wolters led 99 officers and 1,104 enlisted men into East Texas to establish military rule over 600 square miles of oil country. At 6 A.M. he alighted from a train in Kilgore and walked to the new city hall building, in which his staff was setting up his headquarters. Troops moved out to establish encampments, most of them setting up on the Laird farm near Kilgore, others settling down near Overton and Gladewater.

By 7 A.M. military messengers were spreading the word throughout the district that all wells must be shut down by noon. *But neither Governor Sterling's proclamation nor General Wolters' order said a word about curtailing drilling —and while all producing wells were shut down in compliance with the military dictate, unrestrained drilling continued as before.* The rule of capture was still operative.

Production crews were thrown out of work. Most of the local refineries shut down because they had no source of supply other than the giant field. Their employees were laid off. The unemployed and their families were the innocent casualties of martial law.

So were the oil-field poultry. The continuous gas flares from the 1,600 wells had lured millions of insects to doom daily—and the chickens had feasted on toasted June bugs and other tidbits. The *Oil and Gas Journal* soberly took note of this, pointing out that with the flares extinguished, chickens were compelled to "chase their June bugs on the wing or else return to the prosaic ante-petroleum practice of scratching for worms."

Rain fell heavily during the first days of the occupation and the oil field became a sea of slush. Cavalrymen, four in a squad, patrolled the field on horseback day and night. Two National Guard airplanes, piloted by Captains William Ennis and Justin Aldrich, flew over the field daily. The flyers were in communication with headquarters, and were to report any signs of illegal activity. They never saw any, but it is likely that the low-flying planes had a restraining effect. On several occasions the planes were fired upon from the ground, but it was generally accepted that moonshiners were as apt to have been the gunmen as disgruntled oilmen. At least one moonshiner had set up operations *on* the field, disguising his still as a small oil refinery. Daily his trucks, disguised as oil tankers, took off for distant parts with loads of prime booze.

General Wolters dispatched the military band to play in oil-field camps and towns, hoping to keep the people in a good humor. He also issued a press release which said in part: "I trust our wives and sweethearts are busy at home knitting sweaters and making fudge." Since it was only August, and unbearably hot in East Texas, his statement seemed to imply that the guardsmen were in the military district for a long stay. General Wolters was also perturbed about what he called the "painted women" in the district, and he issued an order forbidding the wearing of beach pajamas on the streets, pointing out that the nearest beach was more than a hundred miles away. Beach pajamas, as noted earlier, were the street attire and trademark of the hooker.

The militiamen had their duties, but they also tasted the pleasures of the booming oil-field's fleshpots. They helped the Rangers and the local police and sheriff departments enforce the law, but they closed their eyes to practices as old as man's tenure on earth. Kilgore, for example, had a new jail and a new police force, but somehow no provisions had been made to pay the officers. Consequently, every prostitute in the Kilgore area was arrested once a week and paid a $20 fine, said fine going into a pot to pay the Kilgore police force.

General Wolters' airplanes were stationed at the Tyler airport, long a courting place for the area's younger set.

Since a guard was thrown around the airplanes at night, the young folks were subject to observation by the guardsmen. A group complained, saying they had no other place to romance. General Wolters told them that "war could not bend itself to the whims of the lovelorn."

Four days after the troops arrived, a fire erupted in the city limits of Kilgore. The military said it had been set by arsonists. "Keep a sharp lookout for firebugs . . . and shoot to kill," Wolters ordered. A few hours later three men in an automobile drove into the Kilgore camp and exchanged gunfire with the soldiers. They were not captured, but the shots were the first fired under the occupation. Other fires broke out during the following week, but no one really knew if they were set by angry oilmen, by persons trying to implicate angry oilmen, by children or by vandals. Regardless, the fires were used by those favoring martial law as a reason for its necessity.

General Wolters also decreed that no mass meetings could be held in the military zone. So meetings protesting martial law were held in communities on the fringe. Hundreds gathered in Palestine to hear a speaker shout: "With the National Guard on duty twenty-four hours a day, like watchdogs on a great hoard, the major companies are drilling night and day to sink new wells into the pool that the independents discovered and developed—but martial law places such restrictions on the field that the little fellows no longer can afford to drill."

The small independent operators were hurting, no doubt: with no income from production, they could not afford to drill as could the majors and the large independents. But other "little people" were suffering also, particularly landowners. They had leased their land in anticipation of receiving one-eighth of the oil produced and marketed. Even when they were treated fairly, their one-eighth of oil selling for less than twenty cents a barrel amounted to little. But many of them—particularly the poor and ignorant, both black and white—never had received a fair accounting. Even some major companies were slow in making payments to landowners. The landowners hoped that martial law would save them from bankruptcy, since many of them had spent their royalty in advance.

Many small independents staved off financial disaster by trading their hopes for the future for immediate operating capital. They dealt with larger independents, with well-heeled promoters who had never drilled a well and had no intention of doing so, and even with the enemy, the majors.

A small independent would agree to sell, say, 100,000 barrels of oil in the ground for ten cents a barrel, to be delivered at any time the purchaser requested it. The small independent thereby obtained ready cash to continue drilling and meet current expenses; the purchaser was in a position to wait, and some promoters reaped golden harvests with each price increase on the wildly fluctuating crude market.

A joke illustrating the enmity between the independents and the majors became a segment of oil-field lore. Heartsick and weary from his dealings with the majors, an independent sought solace in a church. As the preacher droned the sermon, the tired independent dozed off. He awoke with a start during the concluding prayer when the preacher said, "Oh, Lord, bless the pure and the humble—"

The independent jumped to his feet. "Hold on there, parson!" he shouted. "Us independents are still in this fight and I want you to put in a word for us!"

While East Texas was still occupied, the Texas Railroad Commission met in Austin on August 25. After hearing wearying arguments for and against proration, for and against martial law, the commission set a new allowable of 225 barrels per day per well with a total of 400,000 barrels daily for the field. The wells were opened on September 5. Governor

Sterling immediately issued a statement saying the well allowable was too high, but the small independents knew that 225 barrels per day was the break-even point for them. Any amount below that would force them to begin selling out to the majors and large independents.

Six days later the well allowable was cut to 185 barrels daily, and three weeks later it was cut to 165. The Commission said the cuts were necessary because new wells in the field had driven production over the 400,000 barrels daily.

Lawsuits by the dozens were filed by both operators and refiners. And the more resourceful among them began surreptitiously to produce above the allowable right under the nose of the troopers.

It was at this point that the term "hot oil" was born. Working undercover for General Wolters, Sergeant W. D. Johnson complained that a sudden rain had made the night chilly as he talked to an operator suspected of producing oil above the allowable. "Lean against that tank there," the operator suggested. "It's hot enough to keep you warm." While writing his report of his investigation, Sergeant Johnson remembered the operator's remark and used the words "hot oil" instead of the cumbersome "oil produced in violation of military orders and in excess of allowables granted by the Texas Railroad Commission." The words stuck, and became a part of the oilman's vocabulary—and the politician's.

Those who ran oil in excess of state allowables did not consider themselves in violation of the law. They considered the setting of allowables unlawful, and for some time, of course, the courts failed to clarify the issue.

Tom C. Patten, a tall, ruggedly handsome oil-field adventurer, told one of the commission agents, "It's my oil, and if I want to drink it, that's none of your damned business."

Patten became something of a living legend when he

Adapted from Ripley's syndicated feature.

drilled three wells on a quarter of an acre in the middle of the main street in London, a hamlet of thirty families. When he laid out the timbers for a derrick for the first well on the small tract he had leased from G. B. Rhodes of Corsicana, a vigilante committee that had been formed to prevent drilling in the streets came during the night and walked away with his derrick timbers, one stick at a time. So Patten called on the services of a friend, one-time Texas Ranger, Clyde Yantzey, widely known as a hard man. Then, with Yantzey at his side, he made it clear he intended to build the derrick and drill. As soon as his well came in, he removed the derrick and, on a plot of ground twenty feet square, sank four telephone poles at the corners, put crossbeams from one to another, and built a one-room house on top of the beams. It became known as the penthouse over an oil gusher and was so noted by Robert Ripley, famous for his *Believe It Or Not* series in hundreds of daily newspapers around the world. People called it the Tower of London on Alibi Hill.

Patten erected corrugated steel around the poles, with the valve connections for the wells in the center of the twenty-foot square. He installed an electric staircase (which he could raise or lower with the push of a button) to the house, about twenty to twenty-five feet above the street level. After his house was built, Patten then drilled two additional wells on his small tract, each capable of flowing more than 10,000 barrels of oil per day.

Patten prepared to produce his oil. On the flow assembly of his first well he hollowed out the main valve so that oil flowed through it whether it was in an open or closed position. From it, he ran a pipe up into his derrick penthouse, where he installed a working valve to control the flow from the well. With the main valve on the well in a closed position —to satisfy commission investigators—Patten could produce all the oil he wanted to sell by manipulating the working valve in the penthouse. Then, with this all done, Patten filed to make the penthouse his homestead under Texas law. This meant no one could legally enter his penthouse without a warrant or his permission.

After the third well on the tract was drilled, Patten ran flow lines from his second and third wells into the penthouse, where all three wells were then controlled by the same flow assembly. The oil from the wells went into two 50,000-barrel tanks and from there to a loading rack on a switching line near the London depot. Patten ran his wells to full capacity before martial law was declared, selling his oil to Crown Central Petroleum Company. The oil was shipped in tank cars to the company's Houston refinery.

Patten would see the tank cars loaded, then drive to Houston to collect in cash for his oil.

The shutdown under martial law made it impossible to ship oil by train or truck without detection. Patten obtained an injunction against the military to protect his home—and the valve—from "search and seizure." Soldiers, however, kept watch on the wells by patrolling outside the property line.

Patten went to see J. D. Wrather, an oil producer and the owner of the Overton Refining Company at Kilgore. Wrather was one of the most controversial men in the field. Like Patten, he believed he should be able to do as he pleased with his own oil, whether it was oil he produced or oil he sold as refined products. He was vilified by many, but upheld by the small independents for whom he had provided an oil market when the majors refused to buy. And he was a no-holds-barred fighter.

Wrather needed oil for his refinery, Patten had some. They made a deal. Since there was no way to accurately gauge how much crude Patten would be running to Wrather's refinery, the men agreed on a flat payment to Patten for crude from the penthouse flow assembly. On dark, rainy nights when the guardsmen were not too vigilant, Patten laid a pipeline from his penthouse to another line running to the refinery. Patten's line was buried several feet beneath the muck.

The oil began to flow. Almost every day Patten would go to Wrather's office. From his desk Wrather would take an old shoe box and count out the money until he filled Patten's brief case.

The military, becoming aware that Wrather was refining crude, suspected Patten as the supplier. A metal detector was brought in from Houston. A search around Patten's property line disclosed the presence of the underground pipeline. A bulldozer ripped out the line and tore it in two.

Patten promptly took a train to Kansas City where he

bought a quarter of a mile of fire hose. He had the metal connections removed and the hose pieces sewed in one length. Back in the field Patten had another ditch dug and the fire hose buried. Once again he began supplying oil to Wrather. Again the military noticed that the refinery was operating. Again the metal detector was brought into play, but this time it failed to disclose anything suspicious. Weeks passed before the guardsmen decided to dig a ditch—completely around Patten's tract if necessary. The fire hose was found and pulled from its resting place.

Undaunted, Patten bought a drugstore across the street from his penthouse. He kept the business running in the front while his crews tore out the flooring of a back room and proceeded to dig a tunnel to the penthouse. The tunnel was fifteen feet below the surface and high enough for a man to stand in it. More fire hose from Kansas City was laid in the tunnel as the pipeline extended from well to drugstore and from the drugstore to the line to the refinery.

It took Patten two months to complete the tunnel and the military and commission investigators four months to find it.

Wrather, meanwhile, sought and received a temporary injunction in federal court which prohibited the military from interfering with the producing of the wells. The military did not obey the injunction, however, and the court found General Wolters in contempt. Patten produced and sold more than a million barrels of oil from the wells below his "homestead." The day that courts finally upheld the state's allowable law, Patten discontinued selling oil above the allowable.

But E. O. Buck, newly hired as an engineer by the Railroad Commission, made a tour around the Overton Refinery one night with another engineer. They found a pipeline throbbing with the passage of oil inside it. Making sure they were not on refinery property, which they had been enjoined not to enter, the engineers called for several trucks used in pumping cement. They tapped the throbbing line—

and began pumping cement into the refinery. That shut down the refinery operations for some weeks, and the facility became known as "the refinery that turned to stone."

Although he was only twenty-nine and just a few years out of Texas A. and M. College, Buck was a brilliant engineer with a wealth of varied oil-field experience. He had arrived in East Texas on the first day of martial law, having completed a pipeline survey from Port Neches to Kilgore for the Atlantic Refining Company. He took a job with the Railroad Commission at $230 a month. Shortly afterwards, the commission sent in 175 new men to help regulate production. Only three had ever seen an oil field. Many were from other departments of the state government; they had been laid off from their old jobs because the state was broke and the departments no longer functioned at full capacity. They were former clerks, state policemen, highway department workers and the like.

The few experienced regulators—called "enforcers" by oilmen—were men who previously had been employed in other fields as umpires for the Central Proration Committee, the voluntary group which had tried—with some success—to limit production in West Texas and elsewhere. There were about a dozen of them. The field was divided into four sections and the men went to work.

The most common device employed by the hot-oil runners was the "by-pass," a valve installed on a pipeline, say, between a well and a storage tank. When the by-pass was closed, the oil from the well would be diverted from the tank to some other destination, perhaps another tank miles away. Buried beneath the mud or in the underbrush, by-passes were difficult to find.

A typical case of hot-oil detection for Buck came at three o'clock one morning. A messenger reported that General Wolters wanted Buck near a road juncture near Joinerville. On arrival, Buck found the general and a large group of oilmen standing by a large gathering system. Lights from parked cars illuminated the scene.

General Wolters said crisply, "Buck, oil's not supposed to be passing through these lines, but one of them"—he pointed to the offending pipe—"is dancing around." He jerked his head at the assembled oilmen. "These gentlemen tell me they don't know where the oil is coming from." He smiled. "I thought perhaps you could help us."

Buck followed the line to a valve hidden in the high grass some yards away. He went back to the group. He told General Wolters he had found the valve. "All I have to do is shut if off and we'll find out where the oil's coming from."

"How?" the general asked.

"Since there's a valve, there must be a pumping station nearby. When I shut that valve, the pressure will get so high that pumping station will fly up in the air like a cork out of a champagne bottle." Buck grinned. "Then we'll know exactly where it's located."

An oilman stepped forward. "Wait a minute, wait a minute," he said. "Let's not do anything hasty . . ."

With this confession, General Wolters extracted a promise from the oilman that he would shut down the pump and pipeline—and obey the orders of the military and the Railroad Commission. The oilman's punishment came in being labeled a cheater by his peers. The commission's orders contained no provision for punishment by fine or imprisonment. Many oilmen, however, feared the threat of bad publicity more than the threat of a fine or jail. The commission did have the right to sue an operator producing hot oil, but it was a tactic seldom used.

Hot oil moved from the field by tanker-trucks and by rail, as well as by pipelines. Tanker-trucks simply loaded at wells which were not closely guarded. Four railroad employees were necessary allies in moving hot oil by rail. The trainmen received $3 each for each tank car they loaded with hot oil and hooked onto a legitimate train.

The major companies were accused of being hot-oil runners time and time again, but never by the Railroad Commission. Whether or not they were running hot oil, it was obvious that they were buying much of it, since the production of hot oil was far too great to be handled by smaller refiners only.

Hot-oil runners were placed in two classifications—those who paid royalty to landowners for both legitimate and hot oil produced, and those who paid royalty only on legitimate oil. Then there were those who stole oil from others. They were legion, and they were skillful. One ring of eighteen men was broken up by a force of Rangers and guardsmen and charged with stealing more than a million barrels of oil from leases near Gladewater. Among those arrested was a Railroad Commission deputy supervisor. It was common knowledge—and easy to understand in such an atmosphere —that some commission employees, making $150 a month, closed their eyes to some hot-oil operations for sums much larger than their paychecks.

Viewed from afar, this great boom in an otherwise stricken land appeared as a gigantic farce. Irony piled upon irony. The commission pondered on Olympus and delivered itself of orders which hardly anyone—even its employees— took seriously. Guardsmen and regulators rode and trudged through mud looking for evidence that would never be used in court. Major companies shouted for proration based on market demand and drilled more wells than all of the independents combined. Legislators conducted investigations of investigations aimed more at destroying old enemies than at arriving at concrete decisions. It was a play in which all of the actors were on the stage at once with each knowing the others' lines, a comedy-crime television series in which the bumbling detective seemed destined to pursue forever his equally bumbling quarry.

★

Martial law had been accepted with surprising eqanimity in the military district. With more humor than animosity, many called the guardsmen "Boy Scouts." For most of the workers and their families the occupation was a novelty, something to discuss and to write about to the folks at home. Merchants and tavern-keepers saw the guardsmen as potential customers as much as keepers of the peace, as did the prostitutes and bootleggers. Most citizens had no clear idea of why the guardsmen had been sent into East Texas; they had seen no evidence of "insurrection" or "rebellion," but they knew there was a lot of crime in the area and a controversy about overproduction. Their attitude seemed to suggest that if Governor Sterling deemed martial law necessary —for whatever reason—well, let him send in the troops.

Some responsible citizens, however, felt certain that the oil controversy would have erupted in violence had not martial law been declared. They were more explicit than the governor. They had heard of threats by landowners who had not been paid royalty by operators. There were rumors that independents who had been refused pipeline service by the majors and other purchasers had threatened to blow up purchasers' installations. No one had complained of being directly threatened, and no such threats had been consummated, but to these thoughtful citizens the aura of violence hung over the field. (Months later such threats, most of them vague, would be used in an attempt to justify continuation of martial law. And months later still, such violence as the threats suggested *would* erupt.)

In any event, shutdown and control of the field was General Wolters' paramount duty, and the first his troops performed. It was obvious that any other duties were secondary. The troops did work with the Rangers and other law enforcement groups. Guardsmen quickly picked up Lone Wolf Gonzaullas' trick of examining a man's palms to determine if he were good or bad. And they joined Rangers in making arrests of oil thieves. Obviously their presence had a deterrent value, for the crime rate in almost every category dropped considerably.

Carl Estes, the firebrand editor of the Tyler newspapers, bristled at the occupation. He had been in the Mayo Clinic in Rochester, Minnesota, when martial law was declared. From his sickbed he had fired editorial salvos denouncing the occupation and warning that the shutdown would produce bread lines. He had called on Humble to head the list of subscribers to finance the feeding of unemployed workers. He returned to East Texas about three weeks after the military district had been established. He sent word to General Wolters that he wanted to confer with him in a room Estes maintained in Longview's Gregg Hotel. The general didn't go, but instead sent his aide, First Lieutenant Charles Perlitz, in civilian life a promising young Houston attorney.

Perlitz, handsome and articulate, was a good front man for General Wolters. Estes—loud, profane, challenging on most occasions—was a polite host. Perlitz answered the editor's soft questions quickly and fully, doing his best to explain the necessity for martial law and what it had accomplished. Estes appeared impressed. Perlitz left him believing the editor had changed his viewpoint, if only slightly.

Estes did temper his editorials after that interview, though he did not cease criticizing martial law. On one occasion he went so far as to defend Governor Sterling personally against the "snake" and "coyote" who had written some handbills which defamed the governor while decrying martial law. Later, he would find some good in the occupation.

Other newspapers around the state, however, endorsed martial law, and so did the Shreveport *Times*. These newspapers became incensed when as the result of a suit brought by J. D. Wrather and his partner, Eugene Constantin, on October 13, three federal judges issued a temporary injunction against General Wolters' enforcement of proration orders. General Wolters refused to abide by the temporary injunction and was declared in contempt of court. Governor Sterling insisted that the court ruling applied to Constantin and Wrather only; General Wolters relaxed, and E. O. Buck, the commission engineer, promptly pumped cement into the Overton refinery. The governor announced that he would prorate the field himself and, in a series of slashes, cut production to seventy-five barrels a day per well.

The newspaper support seemed to have invigorated Governor Sterling. When a few months later the federal court permanently enjoined military enforcement of proration orders, the governor immediately appealed to the United States Supreme Court. And while the injunction decision had also declared martial law itself invalid, he refused to call the troops from the field, maintaining that the decision was in relation to Constantin–Wrather properties only.

So more suits were filed, guardsmen continued to plow through the mud, and on March 31, 1932, oil was discovered in northeast Cherokee County, extending the field into a fifth county and adding, as it were, more fuel to an already raging fire.

Elsewhere in the land, low crude prices had forced the abandonment of 600 marginal oil fields, the operators of which, for the most part, had arrived in East Texas prepared to tussle with the Black Giant.

chapter fifteen

THE REDHEAD
FROM AMARILLO

If 1931 was bad, 1932 was worse. The Railroad Commission issued nineteen orders for the East Texas field in that twelve-month span, and all of them were declared invalid when challenged in court. One weary federal judge accused the commissioners of deliberately writing orders they knew would not stand up under a court fight, and loud amens were shouted from almost every segment of the oil industry. Exceptions to Rule 37—the spacing rule—were still being granted with reckless abandon, and rumors of payoffs to commission personnel for such considerations were part of the daily gossip in the field.

Pat Neff, the commission chairman, resigned in June to become president of Baylor University in Waco. His seat was taken by Ernest Othmer Thompson, who had gained statewide fame as the "fighting mayor of Amarillo." Thompson, however, was unable to lend his fighting spirit to the commission immediately; he ran for election to the seat in the Democratic primaries of July and August and, after winning a six-year term, spent almost the remainder of the year learning something of the oil business.

Governor Sterling also had an election to contend with in 1932. He wanted a second term—a traditional gift of the people—but East Texas hung around his neck like the Ancient Mariner's albatross. His chief opponent was again Miriam (Ma) Ferguson, whom he had defeated in 1930. Among other things, Mrs. Ferguson blamed Sterling for the worldwide depression, but she did not let him or the voters forget that during the special legislative session of 1931, he had sworn to veto an oil conservation bill based on market demand. Mrs. Ferguson said she favored such a bill, and would get it passed if she were elected. It is likely that such a statement gained her more support among the majors and large independents than was ever publicized.

And Sterling still had the troops in East Texas. He circumvented the federal court order by returning control of the field to the Railroad Commission and leaving the guardsmen on duty as "peace officers." It was obvious, however, that the troops' chief concern was the pursuit of hot-oil runners, whom they seldom seemed to catch. There were, however, two incidents of violence in the field in 1932—a commission regulator was peppered with birdshot as he went to inspect a lease, and a Gulf well was set afire by a dynamite blast.

The incidents, however, only pointed up that evidence presented in the governor's behalf at the federal court hearing on martial law had been, at its best, hearsay. Martial law was justified because someone had heard someone say that someone was about ready to "take the law in his own hands" —and the presence of troops had prevented it. That was the evidence. The court had decided that Governor Sterling had sent in the troops simply to prevent what he considered overproduction of oil, and that the "insurrection" and "open rebellion" were subterfuges.

Sterling was widely supported in the military district by businessmen, political and civic leaders, and oilmen who wanted a continuation of martial law. H. L. Hunt, for example, sent Sterling letters and telegrams of support, and Dad Joiner wrote him that he had changed his mind and now favored limiting production of the field in which he no longer owned a part. The East Texas Chamber of Commerce and similar groups encouraged Sterling to continue martial law.

Sterling was a good man, and it would have appeared more in keeping with his nature if he had sent the troops into East Texas with the flat announcement that he was going to shut down the great field and oversee orderly production—and nothing else. He would have kept the support of the East Texans who favored martial law, and would not have lost that of many citizens in other sections of the big state who found intolerable his circumvention of the federal court order.

In any event, Sterling lost the election. From August until the end of his term his influence diminished. But an event outside his control occurred which provided one of the few bright spots in a dismal year. The United States Supreme Court handed down a decision in an Oklahoma case, *Champlin Refining Company* v. *Oklahoma Corporation Commission,* which clearly recognized market demand as an important factor in restraining overproduction. The court found that there was serious overproduction in the country and in Oklahoma's flush pools; that full potential production in the case at issue exceeded transportation and marketing facilities and demand, and therefore caused waste; and that no one, even a producer with ways of disposing of his own oil, had the right to produce so as to cause waste to others. Further, the court reasoned that the effect upon price of limiting production to market demand was "merely incidental."

The decision augured well for passage of a market demand law in Texas. J. Edgar Pew of Sun, a staunch conservationist, began a movement for such legislation by attempting to woo the support of the politically potent independents; Sun raised the price of East Texas crude to $1.10 a barrel, and Pew urged the other major purchasers to follow suit. Some did, but others—notably Humble—refused to go along. Independents, politicians and editorialists took turns blasting Humble, but the company stood fast, arguing that if production were limited to demand, prices would take care of themselves. Even Carl Estes, the Tyler editor, charged that Humble wanted to "break down proration in order to buy . . . cheap oil!"

While this hurricane was blowing, a federal court upheld another injunction against the Railroad Commission's latest proration order. This amounted to lifting the lid on East Texas production. Wells flowed wide open, more were drilled, and the field's high-grade crude went out on the market in a flood that drowned the price per barrel.

The legislature bowed before such a rush of powerful events. After a bitter fight, and by a close vote, a market demand bill was passed on November 12, 1932. Governor Sterling signed it upon passage. The new law included in its definition of waste "the production of crude petroleum oil in excess of transportation or market facilities or reasonable market demand," and it did not prohibit the consideration of economic waste in the regulation of production.

At long last a weapon had been forged—but no ammunition had been provided for it.

On December 12, almost sixteen months after the troops invested East Texas, the United States Supreme Court delivered its decision on martial law. It declared invalid the orders issued by Governor Sterling to enforce proration in the field. "There was no room for doubt that there was no military necessity which, from any point of view, could be taken to justify action by the governor in attempting to limit com-

plainants' oil production," said Chief Justice Charles Evans
Hughes. Like the lower court had done, Justice Hughes criti-
cized Governor Sterling, maintaining that "insurrection"
and "open rebellion" had been subterfuges.

By now General Wolters' headquarters was a large pyra-
midal tent on a gentle rise outside Kilgore. On a dark and
misty day, he waited there, to turn over affairs to a civilian.
He was weary after twenty months of conflict that sometimes bordered on the ludicrous. Many would say he and his
troops had saved the field and the lives of many citizens.
Others would say he was a "ham actor" who loved playing
soldier. Some would recall the joke that made the rounds of
the field: "General, do you believe in martial law?" "Yes—
where?" Still others would remember his warmth and his
wry humor—always present but not always revealed.

He was glad to be leaving. Whatever had been accomplished, he knew that on this very day 100,000 barrels of hot
oil were being produced from the thousands of wells visible
through the open tent flap. Again, the price of oil had fallen.

The replacement arrived. He was Ernest Thompson, the
new member of the Railroad Commission. He was called
"Colonel" Thompson, and there was nothing derisive in the
appellation. He was an authentic World War I hero. He also
was a National Guard officer, but he was here as a civilian.

In Austin he and the other two commissioners had decided that one of them should take up temporary headquarters in the field. Lon Smith and C. V. Terrell were too old
for such an undertaking. They had asked Thompson to take
on the responsibility. It was at least implied that from now
on East Texas decisions would be his to make. Thompson
had grabbed at the chance.

As he entered General Wolters' tent, there was no cere-
mony. As the two men shook hands briefly, Thompson
asked, "Do you have a name for your headquarters, General?"

General Wolters shook his balding head. "Never got

around to naming it officially, but I've thought of naming it hell."

Thompson smiled. "If you have no objections, sir, I'll call it Proration Hill. They both mean the same thing."

The general shrugged and pointed at some papers on a desk. "You just sign that memorandum receipt and you can have your hill and proration, too."

Thompson signed and the general departed. The receipt was for horses and equipment for a few troopers who were to remain in camp to aid Thompson in his first assignment.

Thompson looked out through the tent flap at the forest of derricks. The great field throbbed with life. He rubbed a hand across his fiery red hair. This was his first intimation of the immensity of the field—and the immensity of its problems.

Thompson was forty—a tough, intelligent and ambitious man. He had never suffered a major defeat, even in his boyhood. He had been a successful businessman and lawyer, an outstanding soldier, and possibly the best mayor ever to serve a Texas city. As chief executive of Amarillo, he had forced the utility companies to reduce exorbitant rates by a series of spectacular moves that had gained him nationwide attention. In fighting the telephone company, for example, he had asked citizens to leave their receivers off the hooks, and the citizens had responded almost to a man. Then, since the company's franchise was about to expire, he threatened to have city crews cut down every telephone pole on city property at one minute after the expiration deadline. He had won both that fight and battles with the gas and electric utility companies.

He had not sought the job on the Railroad Commission. Governor Sterling had summoned him to Austin and had asked him to take the seat Pat Neff was vacating. He owned neither oil land nor oil stocks. Of East Texas he knew only what he had read. He had agreed to take the job because it was a public service, because it was challenging, and because he was very ambitious—and not necessarily in the order named.

He already had taken a decisive step. In Austin, before his departure for East Texas, he had voted that the field be shut down until the commission members could gather their wits and sort out the problems. It was, primarily, a defensive gesture, because a commission order of December 10—two days before the end of martial law—had provided an allowable formula based two-thirds on a well's production potential and one-third on acreage and bottom-hole pressure. The inclusion for the first time of bottom-hole pressure in the allowable formula had brought howls of protest from operators on small tracts of land. They favored allocations based on wells only, and they represented a powerful political force.

Thompson knew little about bottom-hole pressure, and his veteran colleagues knew even less. They had included it in the allocation formula because many company engineers had insisted that they do so. And E. O. Buck, the commission's young engineer, had on short acquaintance, bluntly told Thompson that this formula—or one similar—would save hundreds of millions of barrels of oil for mankind's use. Thompson had been more impressed with Buck's earnestness than with the company engineers' charts.

The great field had been as much of a riddle to petroleum engineers as it had been to geologists and geophysicists. What made it tick? What kind of energy forced the oil through the borehole to the surface in such volume and with such power? That was the primary question. Until 1930 the scientists had concentrated their studies almost exclusively on the role gas played as a reservoir energy source—both free gas and gas in solution. Free gas rested like a cap on the crude; gas in solution, of course, was gas dissolved in the crude. Both forced the oil to the point of lowest pressure, the borehole.

But there was no free gas in the East Texas field. And

countless hours of study and experimentation revealed that the crude was undersaturated with gas; therefore, dissolved gas could not be the effective source of the field's energy. Since the field had been opened the gas had been separated from the oil at the surface and flared as a waste product. In other fields, engineers had experimented with injecting the gas back into the reservoir to maintain pressure, but obviously this was not called for in East Texas.

Well, then. Could it be the water? Certainly there was salt water beneath the oil-bearing Woodbine and possibly within it. Had not the great basin been formed by the undulations of a mighty sea? Did not the drill encounter salt water when spun too deeply past the Woodbine? But such water would be static, incapable of exerting pressure on the oil globules, unless it was replenished by surface waters . . .

The Woodbine outcropped at the surface in the Dallas–Fort Worth area, some 150 miles west of the field. Could water entering the Woodbine there flow through it for 150 miles to displace crude produced from the field? Months of study followed. Cores were taken from the field and studied for porosity. The question: How fast can water trickle through these rocks? The answer: Not fast enough to displace the oil now being produced.

Scientists from almost every major company had tried to solve the great mystery. They had shared facts and theories. And it had been the brilliant Ben Lindsly of the Bureau of Mines who had learned that the East Texas crude was undersaturated with gas and thus had eliminated gas in solution as the energy source.

Now it was Humble's exceptionally brilliant research engineers—T. V. Moore, H. D. Wilde, R. J. Schilthuis and William Hurst—who supplied the final answer. They had been studying the flow of fluids, both in the field and with laboratory models. Someone had suggested that they might want to consider the compressibility of water. Scientists knew that water was compressible, but so infinitesimally that no one interested in reservoir energy had heretofore even mentioned it. The Humble men began work on the hypothesis that the natural energy in the East Texas field came from the expansion of water. Their conclusion was this:

> The data on the East Texas field indicate clearly that water drive or water encroachment is by far the most important agency in maintaining the reservoir pressure or in producing the oil. It appears that the water moves into the field by virtue of its own expansion upon reduction of pressure, and not, as in many cases, by flowing through the entire formation from its surface outcrop to the field.

By the time Thompson sat down in General Wolters' canvas chair on Proration Hill, Humble scientists—and others—had concluded that in order to use the water drive most effectively, all wells in the field should be produced at rates that would cause the water to rise uniformly throughout the field as it displaced the oil. They said so—loudly.

Every operator in the field understood the pronouncement: When you take oil from the reservoir in an orderly fashion, it is replaced by water, barrel for barrel, pressure is maintained, and a greater amount of the crude can be recovered from the reservoir; when you produce recklessly and unevenly, water will flood the Woodbine, oil will be lost beyond recovery, and the reservoir's pressure will be dissipated.

Many operators did not believe the verdict. Many others were not disturbed by it; they were in a hurry to make a dollar.

Thompson hadn't made up his mind. But he had decided to try out a plan presented by Buck. If the plan worked, it would help him make decisions he knew he would have to make. Buck had followed the work of the major company scientists with keen interest. Buck was aware that Amerada,

Humble and other companies had by now developed a device to gauge pressure at the bottom of a well—the bottom-hole pressure.

"Let's borrow some of the bottom-hole pressure gauges," Buck had urged Thompson. "Let's take readings on key wells while they're running wide open. Then let's shut down the field—you're already planning that—and take another reading when you open it. If the bottom-hole pressure builds up during the shutdown, we can *prove* to these people that overproduction is damaging the water drive—and we can prove it to the courts."

Thompson and the other commissioners had agreed in Austin to shut down the field for a month. "Make it two weeks, Colonel, or at least announce it's for two weeks," Buck had advised. "These people will stand still for two weeks, but announce it as a month's shutdown and you'll be hit with a hundred injunctions before you get through talking."

Now, sitting at his new desk in the sodden tent, Thompson turned to a young military clerk who was typing some reports. "You know, Sergeant," Thompson said, "somebody is going to have to take hold of this thing right, and it looks like I've been elected . . ."

December 17, 1932, was a raw day, with East Texas experiencing its first heavy snowfall in thirty years. Buck had selected twenty-eight key wells on which to conduct his test; he had been able to borrow just twenty-eight gauges from the majors. The key wells ran from the west, where the water drive originated, to the east, where the field played out against the Sabine Uplift.

The results were even more dramatic than Buck had anticipated. Bottom-hole pressure in wells at the extreme west side of the field was recorded at about 1,400 pounds per square inch. From that point eastward, the pressure dropped steadily to be recorded at about 700 pounds per square inch in wells on the extreme east flank! This figure was even more shocking because Dad Joiner's Daisy Bradford 3—also on the extreme east flank of the field—had shown more than 1,600 pounds per square inch of pressure on a more primitive gauge some two years earlier!

Obviously this drop in pressure was the result of oil being withdrawn faster than it could be replaced in the aquifer by the water drive.

The next day, the eighteenth, Thompson ordered the field shut down. His band of enforcers and the remaining guardsmen went out to oversee the operation. It was a slow process. Some operators argued and threatened lawsuits. Some stood by their wells with shotguns cradled in their arms. Thompson talked himself hoarse, explaining the reasons for the shutdown, and by the night of the nineteenth all but six wells had been closed. These six were enclosed by wire fences, and the operators had turned deaf ears to both orders and entreaties.

The next morning Thompson was awakened by a telephone call from Amarillo. It was his mother. His father was critically ill and she wanted him to come home at once. The elder Thompson broke in on the conversation. "Ernest," he said, "the Dallas *News* says you've got all but six wells closed in. Don't you leave there until they're closed down. Close them yourself, and then come home."

Thompson promised. He called in some Texas Rangers to accompany him to each of the barricaded wells. At each stop he made the flat statement: "Shut 'em down or we'll shut 'em down for you. If you try to use guns, we've got guns, too."

The wells were shut down. Thompson left a Ranger at each one to see that they were not reopened. He returned to his tent headquarters to learn that his father had died soon after the telephone call.

Seventy-two hours after the field was shut down, Buck tested his twenty-eight key wells again. The second half of the reservoir's story was as dramatic as the first. While there was practically no pressure change in wells at the extreme west side of the field, the 700-pound pressure in wells on the extreme east flank had risen to 1,300 pounds!

Buck and other engineers concluded that the field could produce 450,000 barrels of oil daily without endangering the delicate reservoir mechanism. In order to abide by the new market demand law, however, he recommended that the daily allowable be cut to 350,000 barrels.

Thompson then did something that appeared out of character for the bold, imaginative man he had shown himself so far to be. He accepted Buck's recommendation, then issued an order that each well in the field would be allowed to produce 28 barrels daily to conform with it. He ignored acreage and well potential as formula ingredients, though he knew full well that for months the courts had frowned on the old flat per-well allowable.

The allowable was widely ignored, production soared, the price per barrel fell even lower. Landowners and operators began organizing to fight hot oil, and several pipelines were dynamited—violent acts that had been threatened before martial law had been declared. Some of the larger companies, including Humble, employed armed guards to patrol their installations.

And the order based on the per-well formula was declared invalid by a federal court, which also cited the commission for contempt in writing orders that represented "no real variation" from others that had been declared invalid. The Fort Worth *Star-Telegram* editorialized that the commission was a "total failure" and a "standing joke."

Powerful forces among the prorationists began a movement to take the petroleum industry out of the hands of the Railroad Commission. They urged creation of a new state agency, the Oil and Gas Conservation Commission, whose three members would be appointed—not elected, as were the railroad commissioners. They maintained that the railroad commissioners were more interested in votes and favors than in proration.

One of the prime movers for the new agency was H. L. Hunt, the largest independent operator in the field. But Hunt also was beset with a personal problem, one that had captured the interest of all East Texas and of oilmen everywhere.

Dad Joiner had filed a suit against Hunt and his partner, Pete Lake, charging that they had misled him into selling his vast holdings for a pittance! He had asked the court to return his properties to him or to order Hunt and Lake to pay him $15,000,000, which he claimed was the reasonable market value of the leases and improvements, even if not another well was drilled thereon.

chapter sixteen

JOINER VERSUS HUNT

Hunt was confined to his bed in his new Tyler home when he was informed that Joiner intended to file suit against him in a matter of days. Hunt was aware that the statute of limitations on his deal with the old wildcatter would run out at midnight, November 25, 1932. If Joiner wanted to sue, he would have to make his move before the deadline expired. Hunt had less than a week in which to persuade Joiner not to take such action.

He was strapped in a steel brace due to a severe back injury he had suffered in attempting to lift an overturned automobile off the body of a man who had been pinned beneath it. Despite the pain and the handicap of the brace, Hunt had himself moved to a suite in the Baker Hotel in Dallas, where Joiner was living. He saw to it that Joiner was informed of his presence in the hotel and told that he was incapacitated.

Joiner visited Hunt to commiserate with him. Neither

brought up the possibility of a lawsuit. Joiner began dropping in on the ailing man twice a day. Still the possibility of
a lawsuit was not mentioned. During the late afternoon of
November 24, Joiner and Hunt talked for hours. Their conversation ranged over many subjects of mutual interest—all
except the one that was nagging Hunt. Finally Joiner stood
up to leave. Hunt pulled himself from the bed and walked
Joiner to the door. There, Hunt made his decision.

"Mr. Joiner," Hunt said, "I think efforts are being made to
get you to sue me. I hope you don't fall for that."

Joiner reached out and placed his hand on Hunt's shoulder. "My boy," he said softly, "I would never do a thing like
that. I love you too much." And tears came to the old wildcatter's eyes. He turned and walked out of the room. Hunt
returned to his bed, sighing with relief. Now he could relax.

He lay in bed thinking that he had made a smart move in
coming to Dallas. But as he was congratulating himself, he
remembered the tears in Joiner's eyes. Why had the old
wildcatter become so emotional? Suddenly Hunt sat up so
quickly that pain shot through his body. Joiner had cried
because he *was* going to sue!

Hunt had never secured an indebtedness in his life; it was
a basic business principle. He owed a lot of money on thirty-,
sixty- and ninety-day notes. He picked up the telephone and
called his lawyer, J. B. McEntire. He told McEntire to drive
from Tyler to Dallas immediately with several stenographers. Then he called his brother, Sherman Hunt, also in
Tyler, and told him to be certain that McEntire made a
rapid departure.

The next call was to the Continental Supply Company
headquarters in St. Louis. "Send your head credit man down
here right away," Hunt said.

"Why?" asked the Continental executive on the line. "Is
anything wrong?"

"I want to secure my indebtedness. Send him on."

"There's no need for that, Mr. Hunt. We're satisfied with
the condition of the account."

"Send him," said Hunt. "I don't want to discuss it any more."

The executive said he would send the credit man.

McEntire and the stenographers arrived. Then began a mortgage-writing marathon. Before dawn Hunt had written mortgages on his properties in the amount of some $3,-000,000. They had been properly signed by all interested parties, including the Continental credit man.

Special messengers rushed the papers to the Henderson courthouse and filed them as soon as the courthouse doors were opened. An hour later attorney Fred Weeks filed Joiner's suit. Had not Hunt secured his debts, establishing valid prior claims on his properties, Weeks could have placed a lien on Hunt's entire operation in an effort to force a settlement.

On the day the suit was filed, two years after Hunt's acquisition of the Joiner properties, the acreage held 900 producing oil wells!

Joiner's charge was not an unfamiliar one in the oil industry, and the man who made one generally was greeted with friendly jeers, was labeled a "sucker," and was advised to sally forth and find a "sucker" of his own. Scouts who stole or bought valuable information, for example, were regarded as resourceful and daring, and no opprobrium was attached to their activities. But Joiner and Hunt had so captured the public imagination, and their deal had been of such magnitude, that the suit created widespread interest.

It will be recalled that shortly after the Daisy Bradford 3 was completed on October 5, 1930, a series of dry holes was drilled in the vicinity to the north and east. Hunt spudded in a test to the south. He had small hopes for the test, however, and he was almost totally involved in attempting to make a deal for Joiner's leases. This meant he had to spend most of his time in Dallas, where Joiner maintained his office and a hotel residence. Joiner was not feeling well. The Daisy Bradford 3, the eighty acres on which it sat, and the 500 acres he had used in forming his syndicates were being administered by a receiver. He had not gathered his energy to begin drilling operations on his remaining 5,000 acres. Nevertheless, he had repeatedly turned down Hunt's offers for his holdings and had made no counterproposals.

Convinced that a great oil pool lay to the west of the Daisy Bradford 3 on the bulk of Joiner's acres, Hunt had set his scouts to watching a test being drilled less than a mile away to the west. This was the Deep Rock Oil Company's wildcat being sunk by Foster and Jeffries, drilling contractors. Frank Foster was in charge of the operation. The nearer the drill bit got to where the Woodbine should be, the more intensive Hunt made his sales talks to Joiner. Joiner still resisted.

An oil-saturated core was taken from the Deep Rock test shortly after dark on November 26, and at 8:30 P.M. Hunt's chief scout, Charles Hardin, phoned Hunt with the information. Hunt was in the Baker Hotel in Dallas, where he had rented a number of rooms and had been trying to deal with Joiner. Some four hours later Joiner had signed an agreement to accept $30,000 in cash: $24,000 as a down payment for his 5,000 acres, and $6,000 as a down payment on his interest in the discovery well, in the eighty acres surrounding it, and in the 500 acres he had put into his syndicates. The remainder of the $1,335,000 total price he had agreed to accept in promissory notes of $45,000 and oil payments of $1,260,000.

At the time of the suit filing, Hunt still owed Joiner $660,000.

Joiner charged in his suit that Hunt paid Frank Foster some $20,000 to let Hunt's scouts—and no one else—know when the Deep Rock test began showing favorable signs as a producer; that Hunt had kept word of the oil-saturated core from Joiner, and instead had told the old wildcatter that Foster had drilled past the point where the Woodbine was expected without finding the precious sand; and that Hunt had so deprecated the test and the area around it that Joiner had accepted the $30,000 down payment and the involved oil payments when he could have received three to five times Hunt's total offer—much of it in cash—within hours after the rich core became public knowledge. The old wildcatter charged that Hunt had kept him so occupied in his hotel rooms that Joiner had no way of knowing what was occurring in the area around the discovery well.

Hunt denied it all. Certainly Joiner had been "in and out" of his hotel rooms all day of November 25 and 26. But he had kept Joiner up to date on every report he had received on the Deep Rock test. He had told Joiner of the great core taken from the borehole. He had done this to assure Joiner that though he was offering Joiner very little in cash, his future payments in oil looked very good. Indeed, said Hunt, Joiner had agreed to accept this offer hours before Hunt had received information about the core. As for the $20,000 in negotiable notes he had given Frank Foster . . . well, it was for something else.

"What else?" demanded attorney Weeks.

At first, Hunt said he gave Foster the money for a lease which he did not buy because he decided it had no value.

"Still you gave him the same notes, the same amount of notes, without the lease that you were going to give him with the lease, is that right?" Weeks asked.

"Well, he already had the notes," Hunt replied.

"Now, what did you give him those two ten-thousand-dollar notes for?"

"I will say for the lease, and in settlement of claims he was making against me."

Weeks wanted to hear about the claims. First, said Hunt, Foster had claimed that Hunt owed him a commission on a deal that had not been consummated.

"Well, you didn't feel like you owed him a cent for that, did you?" Weeks asked.

"No, none whatever."

"In fact, you knew you didn't, didn't you, Mr. Hunt?"

"I was sure he couldn't collect in court."

"Yes, sir. You didn't value that claim of his at five cents, did you?"

"I didn't value any of his claims as having any merit."

"But you did value the fact that he let your man have that core that night, didn't you?"

"I have always been appreciative of all the information he furnished us from that well and all other wells that he drilled that we had information from."

"Yes, sir. And you paid him that two ten-thousand-dollar notes in token of your appreciation for that service, didn't you?"

"No, sir," Hunt replied.

"All right. Let's go back then. Let's just see what you did give him the two ten-thousand-dollar notes for. What were the other claims he made?"

"Well, he claimed he should have been given some drilling contracts when we had surplus drilling contracts, that we'd given surplus drilling contracts to Sam Cook and other parties when we did have contracts to let. He claimed that."

"Well, now. Let's see about the drilling contracts. How much of those notes did you give him because of the fact he claimed that he ought to have had some drilling contracts?"

"None."

"You didn't give him any notes for that, then?"

"No."

"Well, let's see what other claims he made against you for which you could have given him the twenty thousand."

Hunt explained that Foster had talked with Pete Lake about Foster and Lake buying out Joiner and that because Lake had joined with Hunt in making the deal, Foster thought he was owed something.

"So Mr. Foster claimed that because Lake went in with you and bought that instead of going in with him . . . that you ought to give him about twenty thousand dollars—is that right?"

"No, he asked for seventy-five thousand dollars."

Hunt explained that the $75,000 demanded by Foster was for "all the claims." He said that Foster also claimed a commission on the Hunt–Joiner deal. Under Weeks's questioning Hunt agreed that Foster had done nothing to earn a commission and did not deserve one.

In fact, Hunt said that his employees had been of great help to Foster in drilling the Deep Rock well because they had supplied boiler steam and furnished geological information.

"Instead of you being obligated to Foster for something at the time he showed your man the core, Foster was really under obligation to you for your kindness and favors you had shown him—isn't that right?" Weeks asked.

"Well, I think he would feel under obligation to me, not only for the consideration that had been shown him down in the field, but in other fields."

"Yes, sir. But on top of that—being under obligation to you—on top of that, he made a claim against you for seventy-five thousand dollars—is that right?"

"Yes, sir."

Again and again they went over Foster's claims. Hunt added a new one: Foster thought he was entitled to some financial consideration because he had brought in a well, the Deep Rock, which proved that the area to the west of the Daisy Bradford 3 was productive.

Weeks took another tack, and Hunt said he had paid Foster to keep Foster from suing him and attaching his property. He was particularly concerned that Foster might interfere with the pipeline runs of the Panola Pipeline Company, Hunt's outlet to the market.

Weeks wanted to know how Foster could attach the giant Hunt properties by suing for a $75,000 commission.

"Well, I don't know," said Hunt. "Sometimes they do."

"Didn't you know, as a matter of fact, that he had nothing on earth for which he could file and stop your runs?" Weeks asked.

"No, sir, I didn't."

"Well, how could he do it?"

"I'm not that good a lawyer to say," Hunt said.

Weeks brought out that Hunt had not paid Foster the $20,000 until February 1932, more than a year after the Hunt–Joiner deal. He wanted to know why.

"Well, he never made any claim or insinuated he was claiming anything from me until 1932," Hunt said.

"Did he tell you he wanted to wait to get that money until after he got his Deep Rock affairs pretty well closed, so he would get that money and the Deep Rock people wouldn't get anything?"

"No, sir."

"All these big claims he had against you back there on account of not getting in with Mr. Lake and all these other things—he never hinted any of this up until 1932, is that right?"

"Yes, sir."

"He wasn't a poor boy in 1932, was he?"

"Yes, he said he was in desperate need of money at the time he came to me."

"Did you talk to his partner, Mr. Jeffries, about that core?"

"No, sir."

"Did you know that Frank Foster told Mr. Jeffries that he got the twenty thousand dollars from you as a commission on the Joiner deal and without mentioning any of these other claims you've spoken of?"

"No, sir."

"Well, you don't know how Foster ever earned twenty thousand dollars in commission on the Joiner deal, do you?"

"No."

"You didn't think he was entitled to one cent of the twenty thousand dollars you paid him—is that right?"

"Yes, sir, that's right."

Hunt denied the allegation that he paid Foster for keeping him posted on the Deep Rock well and for letting Hunt's agents have the core. It was generally acknowledged that whether Hunt paid for the information or didn't had little bearing on the case—if case there was.

Street-corner and tavern discussions centered on two points: Hunt said he had kept Joiner informed of the progress on the Deep Rock well, even to the point of telling him about the rich core; Joiner said Hunt had told him nothing but bad news about the well. Joiner's brief maintained that "the defendants secured a number of rooms at the Baker Hotel . . . and invited and caused plaintiff to come to such rooms about the time the defendants thought the Deep Rock well was about to be brought in, which was some two days or more before November 26, 1930, and caused him to remain in said hotel in order that the defendants might keep the plaintiff under their supervision and influence and prevent any other prospective purchasers of said leases from contacting him, and to prevent plaintiff from securing information concerning said well except such information as the defendants desired to give him." Hunt maintained that Joiner had been free to come and go as he desired, that he had made and received telephone calls, and that he had talked freely with persons who visited the hotel rooms.

Charles Hardin, Hunt's scout who had received the core from Foster and had called Hunt with the information, was listed as a Joiner witness, but the sheriff's office returned a Hardin subpoena as not served. It was generally believed that Hardin had told Weeks and Joiner about the $20,000 payment to Foster.

Dad Joiner resolved the case by walking into court with a prepared statement which was accepted by the judge.

At the time I filed this suit it was my information that the cause of action, if any, would be barred by limitation

within one or two days after the date of which it was
filed. For this reason, the suit was filed hurriedly. For
the reason stated, I did not have the opportunity before
filing the suit to investigate certain material facts or
alleged facts which I, as plaintiff, alleged in my petition.
I had no opportunity after these alleged facts were
brought to my attention to determine whether or not
they were true or false.

Since the suit was filed, I have made a thorough investigation and have determined to my satisfaction that the
allegations of fraud in my petition are not true in fact.
After making such investigation, I reached the conclusion, and hold to that conclusion now, that I was not
deceived or defrauded in any manner . . . at the time
the contracts involved in said suit were by me executed.

Had I had an opportunity to make the investigation
before filing the suit, which I have made since it was
filed, the suit never would have been brought. I am now
convinced that the trade . . . was fairly made by all
parties to said agreement, and that no oil-saturated sand
had been recovered from the Deep Rock well prior to
the time said sale was made. I am thoroughly convinced
that at the time the trade was made that the defendants
in this cause, nor neither of them, knew any more about
the condition of the Deep Rock well or its prospects of
making a producing well than I knew myself.

I have read the deposition of H. L. Hunt . . . and I am
convinced that every material statement made in said
deposition by the said defendant, H. L. Hunt, is true,
and I now adopt his said deposition as part of my statement and testimony in this case.

I reiterate that this suit was filed upon a misapprehen-
sion of the true facts, and having now discovered the
facts, I am willing that this Court enter this judgment
denying any relief whatsoever to the plaintiff, and that
this judgment shall ratify and confirm the trade made

. . . and ratify and confirm the three instruments in writing evidencing such trade, as I have heretofore identified, and that this Court shall adjudge all costs accrued in this suit against me, the plaintiff.

Judge R. T. Brown announced that the Court, "after hearing the pleas, the evidence and the arguments of counsel, is of the opinion that the plaintiff should take nothing." He also ratified and confirmed the sales contracts, and said Joiner would have to pay all of the suit costs.

Joiner told reporters that he "had not been held under duress" in Hunt's hotel rooms, as many believed he had alleged. Later the old wildcatter sought out Hunt and asked Hunt to advance him $10,000 in oil payments to pay for attorney Weeks' services.

"I'll let you have three thousand for him," Hunt said. "He lost the case, didn't he?"

Joiner argued. Hunt scolded him for the way he wasted his money. Joiner accepted the $3,000.

The lawsuit in no way impaired the personal relationship between Hunt and Joiner. The mutual liking inspired by their first meeting developed over the months into genuine affection. Joiner had quickly spent the $30,000 down payment and the $45,000 paid on the promissory notes from Hunt. His oil payments sometimes amounted to as much as $50,000 a month, but he always appeared to be pressed for funds. Many times on the twenty-fourth or twenty-fifth day of the month, Joiner would present himself in Hunt's office with a plea for an early payment.

"Boy, could you help me out a little bit?" Joiner would ask. "I wouldn't bother you except that I have some checks out."

Hunt would give him the advance, occasionally accompanied by a lecture. Much of the money Joiner was spending on wildcatting in Northeast and West Texas. The Daisy Bradford 3 was the only well he ever drilled in East Texas; he had found the "ocean of oil" there, and now he was looking elsewhere for another.

Hunt had been the defendant in some 250 lawsuits brought against him as a result of his deal with Joiner. He had not lost a single one. After the deal he had found that of the 5,000 acres involved, he had a good title to only two and a half acres! He had cleared title after title by buying the royalty from the landowner, and having the landowner recognize the lease as valid in the royalty contract. He had welcomed the lawsuits, most of which he had settled out of court for less than $250 each, because they established his lease ownership beyond doubt.

chapter seventeen

"TOUGH TOM" KELLIHER

Colonel Ernest Thompson went to Austin from his tent near Kilgore in March of 1933 when the legislature was considering a bill that would take the oil and gas industry from under Texas Railroad Commission control and create a new Oil and Gas Conservation Commission to oversee the industry. Thompson journeyed in some disrepute because the first proration order he had issued as the man in charge in East Texas had been declared invalid by a federal court. He and the other commissioners also had been cited for contempt.

H. L. Hunt and other powerful independents had allied with the majors in pushing for the new commission, which would be appointed by the governor, the speaker of the House, and the president of the Senate. The bill had majority support in both legislative houses, and Governor Ferguson had indicated her willingness to sign it if it passed. Indeed, she had been accused of helping for-

mulate the bill in order that she might manipulate an ap-
pointive commission.

The bill was to be voted on in the House on the day
Thompson arrived at the capitol. He saw little chance of
changing enough minds in the larger legislative body, so he
took his case to a group of powerful senators who were
supporting the bill. He spoke briefly. "Gentlemen," he said,
"I'm new on this commission. Please give me time to hang
up my hat before you take my job away from me. I believe
we're doing as much as any other commission could do,
particularly an appointed one."

Thompson left the senators believing that his simple plea
had improved his chances of retaining his job. As he prepared to leave the statehouse, he was met by some supporters of the proposed legislation, including Charles F. Roeser
of Fort Worth. Roeser was president of the Texas Oil and
Gas Conservation Association and had been active in the
proration fight since its inception. The men were celebrat-
ing because the House had just passed the new commission
bill. They proceeded to berate Thompson, and one of them
called him a red-headed son of a bitch. Onlookers saw
Thompson's face harden and his fists clench, but with great
effort he managed to restrain himself. He spun on his heel
and walked away.

A short time later Roeser and others accosted Representative Gordon Burns of Huntsville in the lobby of the Stephen
F. Austin Hotel. Burns had voted against the bill. Harsh
words were spoken, and a brutal fight ensued. Burns was
badly hurt. The next day, as the Senate was debating the bill,
Burns was pushed onto the Senate floor in a wheelchair. The
story of his encounter with Roeser and others had spread to
every corner of the statehouse. Burns said not a word. He
was wheeled out of the room—and the Senate killed the bill.
Thompson still had a job.

On his return to East Texas and Proration Hill, Thompson
determined to shut down the field again. There had been
some criticism that Buck's study of the key wells had not

been "scientific enough." He would use that as his reason for closing down the field. Buck could make a more "scientific" survey, and Thompson would use the new evidence to support his next order—one which he intended to be valid.

Thompson reasoned that if he could get a federal court to uphold a commission order—whatever the order's merits—it would firmly establish the commission's constitutional right to issue such orders. After that, he reasoned, he could fight it out with all comers over the arbitrary sections of the order. The federal court was as aware as any politician that the boom was keeping the sick American economy alive. The order must contain nothing that would cause a drop in field employment. If it infuriated anyone, it must be the majors. He would deal with them after the order was upheld.

On March 27, every operator in the field produced his wells wide open for two hours to determine the open-flow potential of each one. The next day they reported the results, under oath, to Thompson's office. On April 6, Thompson ordered the field shut down. Buck conducted another survey. From the open-flow reports and the survey, Buck and other commission engineers arrived at each well's hourly potential. They concluded that the Black Giant could produce more than 100,000,000 barrels of oil per day at full throttle!

Thompson let the word leak out that he intended to reopen the field, with each well being allowed to produce 15 percent of its hourly potential. The total would be 750,000 barrels a day for the field. This was as much as the allowable for the entire state had been before East Texas was discovered. It also was 300,000 barrels a day more than Buck had concluded the field could produce without damaging the reservoir—and it was double the market demand.

Major company engineers argued before the commission, meeting in Austin, that no more than 330,000 barrels a day should be produced. While the hearing was in progress, Farish, the Humble president, expressed his pessimism in a letter to Walter Teagle, board chairman of Jersey Standard. In trying to drill East Texas with some degree of sanity, Farish wrote, Humble had lost several million barrels of oil. Now, said Farish, Humble intended to protect itself by developing its leases at the same rate as the surrounding properties. Farish saw the need for a new commission that was beyond political control or the control of any independent or major company, one which would "deal with proration orders on their merits." Another need was a low price for crude over sufficient time to thoroughly discourage uncontrolled development. This would make the industry aware "that in order to produce at a profit it must have the benefit of the low-cost production that follows control and unit operations." Farish was beginning to feel "that perhaps only the law of tooth and claw" could bring about some order.

Despite Farish's pessimism and the testimony of major company engineers, on April 22 the commission issued the expected order. The larger companies immediately objected that the 750,000-barrel daily field allowable was too large, from the viewpoint of both efficient recovery and the market demand principle, and that the distribution of well allowables according to potential production without any consideration of acreage was almost as discriminatory as the flat per-well basis. Suits were filed at once.

But Thompson had planned well. For the first time in the history of the field, a federal court upheld the commission's order. The court held that the 750,000-barrel daily field allowable was "adequate," and the well potential formula was "more equitable" than the previous ones.

Thompson had succeeded in writing a valid order, one that securely established the commission's constitutional authority, but it is unlikely that he had foreseen the havoc it would wreak. Reservoir pressure in key wells to the west dropped from 1,400 pounds per square inch to 1,200 pounds. Salt water began appearing in volume in wells to the west. Prices skidded: Texaco cut its posted price from seventy-five cents a barrel to ten cents; Gulf withdrew its posted price

completely; and Humble followed the lead. The price structure collapsed. With the high allowable plus the hot oil being
produced, East Texas was producing about half the total U.S.
requirement. Buck told Thompson the reservoir pressure
was dropping at an alarming rate, and warned that if the
pressure dropped below 750 pounds it would be almost impossible to draw the inert oil from the Woodbine even with
pumps. Whatever was left at that point would probably be
lost forever.

Thompson did the only thing he could do. He cut the
allowable. The cut, plus the shock of the price-structure
collapse, made production decline. Hot-oil runners could
find no market even at two cents a barrel. Many operators
shut down their wells rather than sell legitimately for ten
cents a barrel or less.

The majors then established a posted price of twenty-five
cents a barrel. Thompson cut the allowable again . . . and
again. The price of oil rose with each cut in the allowable.
It appeared as if Thompson might be on his way to achieving
some balance, but with each price increase, the hot-oil runners stepped up their operations. It was a vicious circle:
Reduced allowables made for higher crude prices; higher
crude prices made for more hot-oil running.

Thompson went to Austin. At his urging the legislature
enacted a law making it a felony to produce oil beyond the
commission's allowable. Back in the field he installed a system requiring producers to sign an affidavit that oil they
wished to sell had been produced under the commission's
orders. The affidavit had to accompany legal oil at all times.
It was very much like setting the fox to watch the henhouse,
for the hot-oil operators were willing to sign their names—
or any names, including Thompson's—to the affidavits. And
notaries public who would plant their seals on a paper bag
roved the field in droves.

Perhaps nothing would have worked for Thompson at this
point. The chaos created by the high allowable in the April
order had already split the ranks of the conservationists. A

large number had by now given up entirely on state control of the field. The directors of the North Texas Oil and Gas Association petitioned President Franklin D. Roosevelt for federal supervision of the oil industry. The Texas oil industry was in a state of collapse, the directors said, and hundreds of independent operators were facing bankruptcy. A conference of oilmen in Washington also urged federal regulation, even to the point of fixing prices for oil and its products. Governor Ferguson joined the chorus, and many state legislators fell into line.

The cries for help fell pleasantly on most ears in Washington. The new Administration was busily at work revamping the national economy, and the nation ran on oil. The National Industrial Recovery Act was being tailored, and the Administration easily accommodated the oil industry by providing for its regulation in this bill.

Meanwhile, a number of oil shippers had sought to evade Thompson's affidavit rule by billing the product out of state. Thompson refused to let the railroad tank cars move. More than a score of railroad lawyers swooped down on Proration Hill, demanding that the cars be allowed to move. Otherwise, they maintained, their clients would be in violation of Interstate Commerce Commission regulations. Thompson still refused, and the railroads threatened to move the oil over his orders.

Thompson immediately sent a telegram to President Roosevelt stating the problem and suggesting a proclamation banning hot oil from interstate commerce. The next day the President issued an executive order exactly as Thompson had requested it. The hot oil was impounded.

The executive order became the most vital segment of the National Industrial Recovery Act, as it pertained to oil, prohibiting, as it did, the shipment—interstate or abroad—of oil produced contrary to state laws. A Code of Fair Competition for the Petroleum Industry was adopted, and the Petroleum Administrative Board was established to devise and administer regulations.

And in September some fifty federal agents arrived in East Texas to pit the Administration's might and majesty against the Black Giant's hot-oil runners. The contest appeared as one-sided as a lynching.

The federal agents got off to a good start. They stated their business briefly: You can produce hot oil, but you can't get it beyond the state's borders. You may have to drink it, for there's not much of a market for it inside the state. Their presence—combined with the average American's belief that the federal government would spend a million dollars to convict a man for the theft of a postage stamp—briefly slowed down illegal operations. The price of crude rose to a dollar a barrel.

But within a month East Texans had the federal agents catalogued. Most of them were political appointees with no law enforcement experience. And while some were conscientious men trying to perform what they considered an important job, it was apparent that others were as susceptible to influence or bribery as some of the state agents. Production rose to 75,000 barrels a day above the allowable.

The federal agents had introduced a federal affidavit to supplant Thompson's. It was the very heart of the federal enforcement effort, but it became a joke in the field. Oilmen solemnly presented the affidavits so that they could ship crude beyond the state line—and now they were signed with such fictitious names as Frank Roosevelt, Hal Ickes, Ed Hoover and even J. Caesar. Sharpies walked around the field carrying stacks of the affidavits, which they peddled for as little as $4 each.

Thompson had been happy to see the arrival of the federal agents. He was an implacable foe of federal control, but he knew an emergency when he saw one. Still he could not rid himself of a fear that the federal government was eager to render more aid than had been called for. His concern was valid; the Code of Fair Competition envisioned federal authority to fix prices and to control production and refining even down to the level of a single well and a single refining unit. And Harold Ickes, Secretary of the Interior and head of the Petroleum Administrative Board, could not hide his desire to become an oil czar. It was equally clear that Ickes believed the industry should properly be classified as a public utility.

Thompson was a state's rights advocate to the core. That alone would have made him fear what he considered Ickes' socialistic ambitions. But Thompson was a politician, too, and he knew that with Ickes as his target, oilmen of every stripe would support him—and consider him their champion. He took after Ickes with a vengeance. It was well he did; no one else connected with the industry had summoned up the courage and intelligence to do so.

Lobbying against federal encroachment began taking up much of Thompson's time. But he also advocated a measure, which was adopted by the commission, whereby that body began using the United States Bureau of Mines estimates of the consumptive demand for oil in setting its proration totals for the state. This was the first reasonable guideline used in establishing market demand.

Thompson also fought for and gained passage of legislation requiring all refineries in the state to report the source of the oil they processed. The measure was aimed at the scores of East Texas refineries processing hot oil. Some had to shut down, while others were bought by the majors and junked. Many continued to operate, however. Though the law became effective in March 1934, three months later only twenty-three of the sixty-nine refineries which stayed in business were reporting crude sources to the commission with any regularity.

All through the spring and summer of 1934, the East Texas situation grew worse instead of better. Hot-oil production increased monthly with neither state nor federal authorities being able to stem it. And the commission continued to grant exceptions to the spacing rule, thereby adding thousands of unnecessary wells to the field. On June 15, 1934, there were 13,512 wells in the field, 390 were being drilled, and locations for 94 others had been selected. Of the 13,512 wells, the majors and larger independents owned 8,144, the smaller independents 5,368. Humble alone owned 1,422. Gulf, next in rank, owned 826.

In the first flush of the boom, the spacing rule had been largely ignored. Those who went to the trouble to obtain exceptions got them in wholesale batches. But by 1934 more sophisticated methods had been introduced; it now cost from $500 to $1,500 to obtain an exception. It was generally believed that the traffic in exceptions could not have persisted without the tacit approval of the commissioners.

Commissioner Lon A. Smith plunged the commission into deeper disrepute by allowing his attorney son to practice before the commission, representing East Texas interests on matters relating to East Texas problems. And Thompson did his reputation no good by allowing his chief enforcement officer in East Texas, E. N. Stanley, to take a leave of absence to engage in some highly profitable sidelines of the oil industry. Stanley was a captain in Thompson's National Guard unit, a friend from Amarillo days. Thompson had brought him into the field as his "muscle man." Stanley was a powerful man with a whim of iron, disliked by many for his capricious orders.

During his five-month leave of absence, Stanley joined forces with Jerry Sadler, a young lawyer with political ambitions. Like Stanley, Sadler had a reputation as a rough-and-tumble brawler. He was a native of a community called Bug Tussle, and he claimed to be the world's champion snuff dipper. Stanley was a civil engineer, and the new firm was called Stanley-Sadler Engineering, Inc.

The firm dealt largely in two areas. One was "condemned oil," crude produced illegally and impounded by the state. It was sold at a low price to "qualified bidders," whereupon it became legal oil and entered legitimate commerce. The other area was "back allowables." This was prorated oil which had not been taken by pipelines when it was available. The allowables accumulated; when a pipeline connection did become available, the back allowable could be produced and sold. It took influence to be a "qualified bidder" for impounded oil and to acquire permits to produce back allowables.

When Stanley's leave of absence was over, he returned to the state payroll with the new title of field supervisor for East Texas. (Some years later Sadler would win a seat on the commission to add another colorful and controversial chapter to the history of Texas oil regulation.)

Prorationists felt a surge of hope when on March 31, 1934, Thompson and Smith signed an order sending into East Texas as chief administrative officer a veteran and highly regarded commission employee, R. D. (Danny) Parker. The order gave Parker a free hand to reorganize the commission's East Texas division. He was authorized to "employ additional men of his own selection . . . and to release employees now serving said Oil and Gas Division, if in his judgment they are not qualified to perform the duties of the position they now hold."

Parker began cleaning house. But men he fired were almost immediately rehired by the commission and placed back in their positions. Parker, a commission employee since 1909, threw in the towel in two and a half months. He resigned, denouncing all three commissioners as political hacks interested only in furthering their political fortunes and rewarding their friends in the industry. "I never was a 'yes man' and I never will be," Parker told reporters, "and now that it is definitely determined that this is what the commission wants, I step out of the picture, and telegraph advance sympathy to my successor if he assumes his duties determined to stop oil thievery in East Texas, for it cannot be done with the commission's politically constructed enforcement ideas."

The next day Parker was fired. This order was signed by

Thompson and Commissioner Terrell. It said, "R. D. Parker, with a free hand and unhampered by anyone, has had complete control of enforcing the oil laws in East Texas for the last two and one half months. The production of excess oil steadily has increased to an estimated amount of around 100,000 barrels a day since he has had the sole direction of the enforcement of the law and our orders pertaining to that field. It is therefore ordered by the Texas Railroad Commission that the services of R. D. Parker hereby are terminated."

The great field had known no heroes. Strong men, yes, and cunning men. Good men and evil men. But the strongest and cleverest of the best had not been able to impose their character or purpose on the others. The field throbbed on, pouring out its million barrels daily, taking its beating like a brute mule pulling the plow of a sadistic farm hand, itself heroic in the stubborn performance of its unique capability.

And then two champions emerged. One was a twenty-nine-year-old Pennsylvanian, a Yale Law School graduate and teacher who was Assistant Solicitor of the Department of the Interior and a Special Assistant to the Attorney General. He was J. Howard Marshall, the man in charge of the federal legal effort in East Texas.

His first months in charge of the field enforcement had been frustrating ones. Marshall was far from naïve, but his background had not prepared him to cope with the wily and often unprincipled lawyers who had fattened on the Black Giant's blood. But Marshall learned.

A plan for his first successful action came to him as he sat in a Boise, Idaho, courtroom, far from East Texas. Marshall had traced a hot-oil shipment to the Eastern States Refining Company on the Houston ship channel. Gasoline processed from the hot oil had been loaded on two tankers that had steamed through the Panama Canal and berthed on the West Coast, one in Seattle, the other in Portland. Marshall had learned that the hot gasoline had been purchased by the Fletcher Oil Company in Boise. He was in Boise to ask the court to restrain the company from unloading and disposing of the gasoline.

Marshall was alone; several attorneys sat at the defense table. As the hearing was about to begin, a tall, fat, red-faced man entered the courtroom and made his way to the defense table. Marshall recognized him. He called out, loud enough for the judge to hear, "Hey, Big Fish! If I'd had any doubts about this gasoline being hot, your being here would sure as hell erase 'em!"

The big man waved genially and took his seat. He was F. W. Fischer, an attorney known as the Big Fish and commonly called Bull because of his bulk. More than any other man in the field he had kept the hot-oil runners in business. Over the years he had filed more than a hundred injunction suits on behalf of producers and refiners, creating confusion and delaying judicial action while his clients continued to operate at full blast. He was more than a clever lawyer; he was a brilliant student of Constitutional law. Marshall admired him for his ability and liked him for his candor.

As the hearing dragged on, Marshall mused to himself that it was a damned silly business that had brought him to Boise. After the hearing—however it ended—he would have to trace the gasoline back to its origin in the field to determine who had shipped the crude to the Eastern States refinery. Why not reverse the order of proof? Why let the oil leave the field in the first place? The affidavits weren't working, certainly, but some system could be devised to do the job, a method of making the shipper provide proof of the oil's origin before a pint left the field . . .

At the end of the hearing—the judge impounded the gasoline for ninety days so that storage charges ate up anticipated profits—Marshall returned to East Texas with a plan. A Federal Tender Board was established. An oil producer was required to prepare monthly reports giving a detailed history of oil produced and sold in that period. The refiner and purchaser had to file daily reports showing the amount of oil purchased, from whom purchased, the location of the well from which it was produced, and the disposition of the oil. And every oilman of every classification was required to keep his books open for inspection. Failure to comply with any of the regulations was punishable by a fine of not more than $1,000, or imprisonment for not more than six months, or both.

A card for each lease on the field was prepared. When a lease had produced its allowable, a red flag was attached to its card. No one then could obtain a tender for oil from that lease.

It was complicated, but it worked. It worked primarily because of the second champion, also twenty-nine, a blocky, square-faced man with a roguish smile and pale blue eyes. He was a former FBI agent who had worked on the Lindbergh kidnapping case and the Urschel kidnapping case, and had helped in the great manhunts for notorious desperados like John Dillinger and Pretty Boy Floyd. He had helped capture Machine-gun Kelly and had cleaned up a white slave ring in Ohio. His name was Tom Kelliher. The Department of the Interior had hired him away from the FBI.

Kelliher had been working undercover for three months, studying the field and the oilmen. He also had been observing the federal agents. Now he went to Washington and returned with a dozen FBI agents he had lured into his service. This action so angered J. Edgar Hoover, FBI director, that he engaged in a hot quarrel with Harold Ickes and forbade FBI agents to aid Kelliher in any way.

Kelliher's first move was to straighten up his own house. He called the fifty original agents together. He knew which ones were honest and competent, and which ones weren't; but the slate was clean as of this moment, he told them. The tenders would not be a field joke as the affidavits had been.

No oil bound for out of state, he said, would leave the field without a federal tender. The incompetent would be fired; the crooked would be arrested and prosecuted. He made this point clear shortly thereafter by arresting two agents and obtaining indictments against them on charges of accepting bribes.

Federal hot-oil regulation was based on Subsection 9(c) of the National Industrial Recovery Act, which forbade transportation of hot oil in interstate and foreign commerce. It, like the regulations inspired by J. Howard Marshall, provided that a violation was punishable by a fine of not more than $1,000, or imprisonment for not more than six months, or both. With these weapons, Kelliher and his twelve new investigators went to work.

They got evidence and made arrests. They so loaded the docket of Federal Judge Randolph Bryant that the judge could not handle the caseload. The accused took the easy road, pleading guilty and paying fines. The $1,000 fines were filling the federal coffer, but they were not stopping the hot-oil trade.

Kelliher was a graduate of Boston College and Georgetown University Law School. He went as a lawyer to see Judge Bryant. The judge was a staunch believer in individual rights, but he had presided over so many injunction suits in the past three years that he had learned much about the great oil field and the necessity to limit its production. After the conference, Kelliher slowed down arrests and instead began seeking injunctions against hot-oil runners. When the evidence sufficed, Judge Bryant would enjoin the oilman to run no more hot oil or be held in contempt of court. Judge Bryant let it be known that hot-oil runners held in contempt would be jailed, not fined.

Hot-oil production dropped from 100,000 barrels to about 30,000 barrels daily. Unable to obtain enough excess oil, field refineries began to shut down. Shots were fired at Kelliher on dark nights, and deputy U.S. marshals were assigned to guard him around the clock. Only the diehard hot-oil runners were still operating, and some resisted federal agents who wanted to inspect their property or their books.

Kelliher met with four independent oilmen who were conservation supporters: H. L. Hunt, Wirt Franklin, Craig Cullinan and George Heyer. It was proposed that a company of Texas Rangers be brought in to aid Kelliher and his twelve investigators. But the state was still broke; Rangers had been laid off, and a company could not be sent to East Texas.

In some manner, the four oilmen made secret arrangements to pay the salaries and upkeep of a Ranger group. Thirty experienced lawmen under Captain Jim Shown arrived at the field and presented themselves to Kelliher. They were the toughest men Kelliher ever had met. Kelliher sent them out to aid investigators.

By November there was only a trickle of hot oil being produced—perhaps 10,000 barrels a day. Forty-five field refineries had gone out of business. Only a handful were left, those who could compete with the large refineries for legitimate oil. The Federal Tender Board was functioning; it was no longer a joke. It appeared that by simply enforcing the law without fear or favor, Tom Kelliher had performed a miracle.

He also had won the grudging admiration of the Big Fish, whose clients were suffering. But Fischer had no intention of letting Kelliher and Marshall win the day, if he could help it. With all of his skill and knowledge he was preparing an ambitious assault on Harold Ickes, President Roosevelt and the National Industrial Recovery Act itself. If he could prevail, his clients would be back in business—and the name of Fletcher Whitfield Fischer, self-taught, small-town lawyer, would be emblazoned across the pages of the country's legal history.

chapter eighteen

"THE BIG FISH"

The Big Fish went to Washington in December 1934, pausing en route in Dallas long enough to be fitted for a cutaway coat, striped trousers and a wing collar—his formal suit for a U. S. Supreme Court appearance. It was a concession to custom Fischer was happy to make. He had dreamed of such an appearance when he had held a book in one hand and the handle of a six-mule plow in the other as he had simultaneously studied law and broken sod in the Oklahoma Territory.

He had passed his bar examination in Oklahoma City, and was at the head of the group of young men who had not studied in law schools but had read the law as he had done. He had become a railroad lawyer in Oklahoma, but had moved to Wichita Falls, Texas, in 1919 when the oil fields in that area of the state had come roaring in. The Depression —and the East Texas field—had almost wrecked Wichita Falls, so Fischer had hurried to East Texas with the other boomers. He had rented an office above a service station in

Tyler—and he had prospered. He had become the best-known and most successful attorney in the great field.

Now, at forty-six, his speech was full of the ungrammatical but image-provoking similes and metaphors of the oil patch. He was cunning, a trickster who played the country-boy role to the hilt. But state and federal attorneys, and counsels for the major oil companies had learned to their sorrow that his knowledge of the law was sound—and that rural juries and sophisticated judges alike were susceptible to his appeals.

Fischer's clients for this Supreme Court hearing were the Panama Refining Company and A. F. Anding, a producer. But if Fischer won the case, his scores of other clients in East Texas would be victors also, for he was attacking the validity of Subsection 9(c) of the National Industrial Recovery Act and, in effect, the very Act itself.

Fischer had initiated the suit in the court of Judge Randolph Bryant. Both Panama and Anding had refused to comply with government regulations, and criminal charges had been filed against them. Over their objections, federal agents had gone onto their property, had gauged tanks and had dug up pipelines in a thorough investigation.

Fischer had asked that agents be restrained from prosecuting his clients for their refusal to furnish verified daily reports as to the production, sale and disposition of crude by Anding, and the purchase, transportation, storage, refining and disposition of refined products by Panama. He also had attacked the regulation which required extensive book-keeping by his clients for inspection by government agents. And he had attacked the validity of Subsection 9(c), claiming Congress had delegated its power to the President in violation of the Constitution.

Judge Bryant had granted the injunctive relief, holding with Fischer that Panama and Anding were not engaged in interstate commerce and therefore not subject to the operation of Subsection 9(c); thus the regulations were not enforceable against them. And he concluded, therefore, that it was unnecessary for him to pass upon the validity of Subsection 9(c).

The United States Circuit Court of Appeals in New Orleans, to which the case was taken by the government, strongly upheld the validity of the subsection, however, while it reversed Judge Bryant's decision. Judge Samuel H. Sibley, who formulated the court opinion, wrote: "The [Petroleum] Code is a novelty in legislation. Its making was not a delegation by Congress of a power of legislation."

While hearing the Panama case, the circuit court also heard the appeal of a similar case involving the Amazon Petroleum Corporation. The two cases were consolidated when Fischer decided to make an appeal to the Supreme Court.

The Court set aside two days for oral arguments, December 10 and 11. Fischer was impressed but not subdued when he entered the courtroom; it was a quiet, dignified chamber, unlike the rough, noisy arenas of East Texas. Just before the hearing began, Fischer observed what he considered a good omen: Justice Louis Brandeis picked up Fischer's slender, twelve-page brief and measured it against the government's thick two-volume brief.

The government's case was presented by Harold M. Stephens, an outstanding assistant in the Department of Justice. He spoke almost the entire hearing day of December 10. He was followed by Jimmy Sayes, a Texas lawyer representing the Amazon Petroleum Corporation. Sayes had agreed to say only a few words in opening for the petitioners; he had not wanted to be a mere spectator at what might be a landmark case.

Nevertheless, when Stephens and Sayes had spoken, it was 3:45 P.M.—and the Court recessed at 4 P.M. Fischer quickly made up his mind that he wouldn't start his main argument only to be cut off fifteen minutes later and have to continue the next day. He felt that if his presentation was broken up, it would lose much of its effectiveness.

He proceeded, therefore, to surprise the Justices with a colorful description of the "viciousness of all these new regulations with regard to the liberty of our individual citizens who are being ruled by bureaucratic edict."

And then he astounded the Court with the story of J. W. Smith. Smith was unique, the only oil operator ever to spend time in jail on a hot-oil charge. Smith had been charged with violating Section 4, Article 3 of the Petroleum Code, which prohibited the production of oil in excess of state limitations. His bond had been set so high that he had spent several days in jail before the bond was made.

But, said Fischer, Smith had been jailed illegally. And he explained: The Petroleum Code had been drafted in the summer of 1933. President Roosevelt had signed the executive order making the code into law on September 13, 1933. But by some unexplained oversight, the order signed by the President did not contain the penalty clause for violation of Section 4, Article 3. Six months later Smith was arrested and jailed under the provisions of the original draft which, as Fischer said, "did not happen to be the exact instrument sanctified by the Presidential signature!"

Justice Brandeis wanted to know how Fischer was aware of this since no one else apparently was. Fischer did not answer directly. He said, "The reason so few people know that the President, in effect, nullified the code lays with this printing outfit in Cleveland, the outfit that prints for free all sorts of laws and government edicts and sends them around gratis to lawyers and judges all over the country.

"Now, when they got this Presidential order on the Petroleum Code, they discovered right away that the penalty clause was missing. So they figured it had been left out by mistake, and they went to the draft copy they had printed and sent out and got the wording of the penalty clause and put it in the Presidential order. That act of printing the penalty clause, of course, don't make it legal. It just compounds the error."

There was a long silence. Then Justice Brandeis spoke. "Where, then, is the respository of the law?"

"Right here in my hip pocket," Fischer said, and he whipped a battered copy of the code from beneath the tails of his cutaway coat. He waved it gently. "This particular copy came from the hip pocket of a federal enforcer who wandered around where he wasn't supposed to be. That's the only place an oilman can find one."

Fischer told the Court that he had attempted to a get a correct copy of the code—the way it had been signed by the President—but had not been able to do so, either from the White House or the Petroleum Administrative Board. He was implying that federal agents were enforcing a law they knew to be invalid.

Stephens, the assistant attorney general, spoke up, saying that Fischer should have gone to the National Recovery Administration to get a correct copy of the law.

Said Justice Brandeis: "The attorney general might have tried the Department of Justice to get a correct copy." Then Smith wouldn't have been arrested by mistake, the Justice was saying.

That ended the first-day session. J. W. Smith and the missing penalty clause had no true connection with the case at hand, but Fischer had created an atmosphere to his liking. It was one he exploited the following day.

Fischer was aware that the National Industrial Recovery Act was emergency legislation and that it was scheduled to expire in several months, but that a movement already had begun to continue it for two more years. So as he opened his argument before the Court, he went back to the days of ancient Rome. A dictator rose up before the Roman Senate one day, he said, and asked the Senators to give him two years of unrestricted power. After two years, the dictator returned and asked for two more years since his program was not yet completed. "And when that two years was up, he went back and *demanded* ten years of unrestricted power, and when that was up he told the Senate he didn't need the Senate any more.

"I figure the NRA was started in the same manner. It's only supposed to run two years, but at this very moment there's a proposal before Congress to extend it for another two years. After ten years of NRA, a dictatorial president could very well tell Congress he didn't need it any more, and send it home."

East Texas, he said, was being run in a dictatorial manner by federal agents with powers not properly delegated to them. He argued three major points. First, he said, Subsection 9(c) of the National Industrial Recovery Act was an unconstitutional delegation of legislative power to the President because it did not manifest the policy or will of Congress upon the question of the prohibition of the movement in interstate and foreign commerce of petroleum or its products produced or withdrawn from storage in violation of a state law or regulation. Whether the commodities were to move freely in such commerce, or to be prohibited from so doing, was left entirely to the discretion and will of the President.

"The Act, when it left the hands of Congress, was not full and complete and capable of enforcement as its [Congress'] Act, but remained a mere nullity until life was breathed into it from an unconstitutional source by the President exercising his discretion to prohibit the movement in commerce of the commodities referred to. If he had elected *not* to prohibit the movement of such commodities in commerce, such commodities would still be free and untrammeled subjects of interstate and foreign commerce, notwithstanding the act of Congress.

"It therefore follows that the prohibition of the movement of these commodities in interstate commerce results from the discretion of the President and, consequently, the Act of Congress authorizing the President to effect such prohibition in commerce at his discretion is repugnant to Article 1, Section 1 of the Constitution, and is void."

But even if Subsection 9(c) were valid the regulations promulgated to enforce it were not authorized by it, Fischer argued. Nor were the regulations enforceable against his clients. His clients were engaged only in intrastate commerce; thus they were not included within the terms or reasonable intendment of the Act, and the things required of them by the regulations were not required by the Act. "Therefore, the effect of the regulations is to extend the scope of the Act beyond its terms, which is legislation and not administration, and they are therefore void."

This brought him to his third point. "Conceding that the attacked regulations are authorized by Section 10(a) of the National Industrial Recovery Act, wherein it is provided that the President is authorized to make all rules and regulations necessary to carry out the purposes of the Act, yet a general grant of authority to make rules and regulations for the purpose of carrying out the terms of an act is not sufficient authority to subject one to criminal procedure for violating such a regulation."

He concluded: "It is, therefore, respectfully submitted that this case is one calling for the exercise by this Court of its supervisory powers, in order that the petitioners may have the relief awarded them by the district court, and that to such an end a writ of certiorari should be granted, and this Court should review the decision of the United States Circuit Court of Appeals and finally reverse it."

The Supreme Court did just that. In an opinion handed down on January 7, 1935, Subsection 9(c) was declared unconstitutional. Several months later in another case, the Court used the Panama case as a precedent and declared the entire NIRA structure unconstitutional.

The Panama decision was voiced by Chief Justice Charles Evans Hughes, and he had few kind words for Congress. To the government's contention that stringent regulation of

the oil industry had become necessary, Justice Hughes
wrote:

It is no answer to insist that deleterious consequences
follow the transportation of "hot oil"—oil exceeding
state allowances. The Congress did not prohibit that
transportation. The Congress did not undertake to say
that the transportation of that oil was unfair competition. The President was not required to ascertain and
proclaim the conditions prevailing in the industry
which made the prohibition necessary. The Congress
left the matter to the President without standard or
rule, to be dealt with as he pleased.

The effort by ingenious and diligent construction to
supply a criterion still permits such a breadth of authorized action as essentially to commit to the President the
functions of a legislature rather than those of an executive or administrative officer executing a declared legislative policy.

The question whether such a delegation of legislative
power is permitted by the Constitution is not answered
by the argument that it should be assumed that the
President has acted and will act for what he believes to
be the public good.

The point is not one of motives but of constitutional
authority, for which the best of motives is not a substitute.

Down in East Texas word of the decision spread from
town to town, lease to lease. Justice Hughes' fine language
was translated into one oil-field sentence: "The Big Fish has
kicked hell out of the government, so cock back your valves
and let 'er rip!"

The day after the decision was rendered, General Hugh
Johnson, the NRA administrator, asked Congress for legislation that would meet the requirements implicit in the Su-

preme Court verdict. J. Howard Marshall received a call
from Tom Connally, the senior Senator from Texas. "Young
man, can you write a law that's constitutional and will stop
this hot-oil racket?" Connally asked.

"Yes, sir."

"How long will it take?"

"Give me forty-eight hours."

"Let me have it when it's ready," Connally said.

Two days later Marshall gave Connally his draft. Meanwhile, Senator Albert Gore of Oklahoma had prepared a
proposal of his own. The two drafts were consolidated and
presented to the Senate on January 18. On February 16, the
Connally Hot Oil Act was signed into law by the President.

The Big Fish advised all in the field who would listen:
"You've had a good run for your money. Now it's time to get
off the gravy train." It *had* been a good run. An estimated
100,000,000 barrels of hot oil had flowed from the Black
Giant and into commerce.

The new law, prohibiting the transportation of hot oil in
interstate commerce, was more stringent and more detailed
than Subsection 9(c) and its accompanying regulations had
been. And its penalty clause was sterner, providing for a fine
not to exceed $2,000, or imprisonment not to exceed six
months, or both. Tender boards were authorized to conduct
investigations with witnesses obliged to testify under oath.

The new law had a salutory effect on the Texas Legislature
and the Railroad Commission. The market demand law was
strengthened in 1935, and other legislation provided for better regulation of intrastate oil movement. On June 18, 1935,
the commission issued a new statewide tender order putting
into effect regulations for the entire state relating to the
transporting, storage and handling of crude and its products.

Commissioner Thompson had earned the plaudits of almost every oilman for his fight against Harold Ickes and federal control of the industry. Ickes was still in charge of the federal regulation program, but he apparently no longer entertained hopes of directing the oil industry as his private domain. On many a public occasion, Thompson had laid bare Ickes's ambition and his ignorance of the industry he wished to control. Thompson had made it clear from the outset that federal control should be limited to the prohibition of hot oil in interstate commerce. That he considered a federal duty. Anything beyond that point was an invasion of state's rights. He had won his point.

Thompson also had been instrumental in the formation and direction of the Interstate Oil and Gas Compact Commission, an association of oil states whose representatives met to consider and advise on conservation matters. The commission had no legislative enforcement powers, but it developed a tremendous influence on both the industry and the public.

This impressive body of law and regulation had been more than four stormy years in the making. It had been prompted by the activities in one oil field, the Black Giant. Now for the first time in history, there was some balance in the oil industry. The days of boom and bust were over. There could be no more production orgies that left ghost towns and crippled reservoirs in their wake.

But while the body of law and regulation had resulted from the excesses in East Texas, it was far more effective in other fields, old and new, than in the great field. East Texas was simply too big to be policed effectively, even if the enforcing agents had been completely dedicated servants of the law—which some of them were not. It was too rich a prize for politicians to ignore, and the men and companies that held title to East Texas were by now the wealthiest and most politically potent in the state.

Against the best engineering and geological advice available, the commission continued to grant exceptions to Rule 37, the spacing rule. Men high in government and those with connections with men high in government made fortunes dealing in confiscated oil. Stories of such activities abounded—and were often accepted as true.

The producing of hot oil, however, had slowed down to a trickle. This was because Tom Kelliher had remained in the field as chief enforcer for the federal program. At times Kelliher was accused of taking the law into his hands, and he never denied it. The charge was repeated by Harold Ickes, his superior, when Kelliher stepped on some large and tender toes.

At two o'clock one Sunday morning, Kelliher and his agents opened an investigation that resulted in his leaving both the field and federal service. His report to his superiors said he and his investigators had found hot oil flowing from a large lease into a field refinery. But the oil was not flowing into the refinery tanks, as Kelliher had suspected; it was being by-passed into a large pipeline, the report said, and thence across the Louisiana border into two 80,000-barrel tanks owned by a major oil company.

Kelliher confiscated the oil, shut down the refinery, and took all of the employees on duty into his office for questioning. Almost immediately the Gregg County sheriff and a number of deputies arrived at Kelliher's office with writs of habeas corpus for the employees. Kelliher explained to the sheriff that he only wanted time to question the men. The sheriff shook his head; he wanted the employees turned over to him right now. No, said Kelliher, the writs couldn't be served on federal property.

"I want 'em," the sheriff said stubbornly.

"You'll have to take them," Kelliher said.

A long minute passed. The sheriff and his deputies were

on one side of the room, Kelliher and a group of Rangers on
the other. The sheriff, having weighed the eventualities,
made his decision. He turned, beckoned to his men, and
walked out of the door.

Kelliher sent in a report, but before he could proceed with
filing charges he was ordered to Washington by Ickes. Kelliher was being too high-handed, Ickes said. He was antago-
nizing legitimate operators. Other agents thought he was
too harsh. Perhaps he needed a change. For example, said
Ickes, at this moment he could offer Kelliher a choice of two
jobs in the Department of the Interior: Supervisor of the
Shenandoah National Park or Curator of the Jefferson
Memorial in St. Louis . . .

Kelliher told Ickes he would let him know.

In less than a week someone issued a press release in
Washington announcing that Ickes had fired Kelliher. Reporters began calling Kelliher, asking for his comment, if
any. They also began calling Ickes. Sarah McClendon of the
Tyler *Morning Telegraph* brought Kelliher a copy of the
release. Kelliher read it, and called Ickes. Ickes denied issuing the release, saying one of Kelliher's enemies in the department must have done it. Ickes promised that he would
rectify the error. Soon a retraction was on the wire service
lines. Some Texas editors who detested Ickes gleefully
printed the firing statement and the retraction side by side.

Kelliher resigned and went to work for Tidewater Associated Oil Company. Editorialists across the state blistered
Ickes, none so well as Sarah McClendon. In an editorial, she
wrote:

> [Before Kelliher's arrival in East Texas] some mysteri-
> ous things were happening. People charged with re-
> sponsibility for enforcing the law talked vaguely of not
> having authority. They indulged freely in the ancient
> pastime of passing the buck. When it came right down
> to the exact point of enforcing the law, nobody seemed

to have any idea of who was supposed to act or how to act.

Kelliher removed these doubts. He went into action with sweeping, searching investigations. He got results. He could show where the body was buried. He could show the time, place and the man.

That his resignation was brought about by shameless interference and inexcusable lack of support by his own superiors is a conclusion reached by almost every honest oil operator.

This resignation leaves a big black question mark for Secretary Ickes to remove. It leaves an open question as to his sincerity in dealing with the East Texas oil problem.

Kelliher was disliked, she wrote, and it was to his credit.

The very fact that he had these common night-prowling bandit-barons howling for his head is proof sufficient that he was serving the public welfare as no other man has served it in the history of the East Texas oil industry.

In the light of these facts, therefore, there is no other conclusion than that he has been betrayed, he has been delivered into the hands of the Philistines, that he has been sacrificed at a critical hour in this industry's history —sacrificed like Nero fed the Christians to the lions.

Does this mean that we are going back to the old system—to the old slipshod, hit-or-miss, fast-and-loose, soft-soaping and money-palming system?

If it does, it means goodbye to hopes for regulating the oil industry.

If it does, it means that Secretary Ickes is woefully lacking in understanding of this industry's needs; it means that our great government bows in shame to the might and power of unpatriotic, unprincipled people who wish to ravish our public possessions as fiendishly as wolves raiding a meadow of sheep.

It means that the state will continue to be robbed by oil pirates, that our school children will be robbed of millions of dollars by men who have no interest whatsoever in the public good.

It means return to the Law of the Wolf—the Law of Fang and Claw.

It is a dismal, disappointing situation . . .

Sarah McClendon's direst fears did not come true. Kelliher had builded well his organization, and his successor did not destroy it. Further, the scorn that had been heaped on Ickes served to deter any flagrant political efforts to tamper with the federal enforcement program. And Miss McClendon kept a vigilant eye open for "night-prowling bandit-barons" from her post in Tyler and later from Washington when she moved there to become a columnist of note.

The Black Giant, it appeared, was coming of age.

3

THE
AFTERMATH

chapter nineteen

BOUNDARIES
AND BARRISTERS

"**I**f you want a successful gathering of long-lost kinfolks, just manage to find oil on the old homestead. They will come out from under logs, down trees, from out of the blue and down every road and byway, but they'll all get there—even some nobody ever suspected were kinfolks . . ."

So spake District Judge R. T. Brown, the East Texas jurist who became a legend in his own time. Judge Brown was weary after a long day in his Henderson courtroom when he spoke so disparagingly of his fellow-man. Ordinarily his wry humor and his gentle nature allowed him to observe the antics of litigants with a mellow if somewhat rueful eye.

It is likely that Judge Bob presided over more cases in a short span of time than any comparable jurist in American history. The boom saw to that; it produced thousands upon thousands of lawsuits, and many of them were heard and settled by Judge Brown or juries in his court. It also produced a tremendous East Texas attorney population. There were less than 50 practicing lawyers in the communities in and around the great field before it was discovered. Within three months after the Daisy Bradford 3 came in, there were 150 lawyers in Longview alone. Next to oil, "lawyering" was East Texas' biggest business. Sooner or later, each of them, shyster or dedicated advocate, walked into Judge Brown's untidy courtroom. The judge one time grumbled to Edwin Lacy, a native East Texas attorney, that so many lawyers were trying to get his ear that he was hard-pressed to separate the plaintiffs from the defendants.

Judge Brown had gained attention outside East Texas when he had refused to place Dad Joiner's properties in receivership, explaining that "when it takes a man three and a half years to find a baby he ought to be able to rock it for a while." He also sat in judgment when Joiner sued H. L. Hunt for the return of his leases. And he was on the bench when Miss Stella Sands, a Dallas spinster, sued the old wildcatter for half of his assets, alleging she was his silent partner.

Miss Sands, an attractive woman of middle years, came to court with a ribbon-bound packet of letters Joiner had written in the lean years before the Daisy Bradford 3 spewed oil. The letters constituted the whole of her evidence. In a proud voice that never wavered, she read them into the record. They were interesting letters, and the women in the courtroom found them beautifully sentimental. Joiner sat with bowed head, listening to the words he had written.

Miss Sands had occasionally been ill during the time Joiner had written her, and the letters were full of his chin-up philosophy. There were accounts of small successes and many failures. He wrote fascinating details of his activities and his estimates of people he had met as he made his way across Rusk County. And there were requests for money— "just another $200." She was to get leases for her help in finding the "ocean of oil." The letters were a sharing of his life.

Joiner offered no defense. Judge Brown ruled that Miss

Sands' evidence, all taken as true, was insufficient to establish a partnership interest in Joiner's property or money. But
he ordered Joiner to pay her $20,000 for the leases he had
promised her and not delivered. The old wildcatter tearfully
assured Judge Brown that he had always "wanted to give her
something." "See to it then," Judge Brown snapped.

On September 8, 1933, Joiner and his secretary, Dea England, went to Juárez, Mexico, where Joiner divorced his wife
Lydia and married Miss England. Dad and Lydia Joiner had
been married fifty-two years and were the parents of seven
grown children, two sons and five daughters. The new Mrs.
Joiner was twenty-five; Dad was seventy-three. The marriage was not completely surprising to Joiner's closest associates. He and Lydia had grown away from each other. He had
left her behind in Ardmore in 1925 when he moved to Dallas
to prepare for his East Texas venture. From then on, they
had seen each other only briefly and after long intervals.
Dea England had come to work with him in 1927. She had
suffered with him, prayed with him, fought off his enemies
and placated creditors. She was no young woman marrying
an old man for his money. Indeed, she knew she would see
precious little of it, for Joiner was breaking the cardinal rule
of wildcatting; he was using his own money instead of promoting money from others as he sought another ocean of oil
from northeastern Texas to the far western reaches of the
state.

The marriage set off a series of lawsuits inside the Joiner
family. Each member of the family, it appeared, wanted to
secure what he considered his share of Joiner's money. Most
of the suits were settled out of court, chiefly because of H.
L. Hunt's soothing influence on the family members.

Hunt, of course, was involved in more litigation than any
other individual in the field. He rarely filed a suit, however.
One of the few he filed was against Gulf. It was a strange suit,
revealing more about the condition of Texas jurisprudence
than about the litigants. Involved was the 960-acre farm of

George Turner. Joiner, Walter Tucker and another associate had obtained a lease on the property. Tucker and the other men had sold their thirds to Continental and Magnolia. Hunt therefore had received only 320 acres of the lease in his deal with Joiner.

Hunt, Continental and Magnolia were conferring as to who should drill the entire 960 acres for all three parties when word was received that Gulf was already drilling the property. Gulf had decided that Joiner's original deal with Turner was illegal and void because Joiner had paid lease rental not in cash but with certificates in his syndicate. Gulf had obtained what was called a "top lease" on the property.

A jury in Judge Brown's court found in favor of Hunt and his associates. The state Court of Civil Appeals affirmed the judgment. Gulf took the case to the State Supreme Court. That court refused jurisdiction, which gave the verdict to Hunt.

By now Gulf had drilled up much of the 960 acres, but proceeds from the oil sold were to be impounded. Gulf refused to surrender. It employed a former Supreme Court judge to handle the case. He prevailed upon the Supreme Court to take the case and pass on one feature of it—could a rental be paid in any way other than in cash, as called for in a standard lease? Years passed before the Court rendered a verdict—and this time it was in favor of Gulf.

Hunt's chief attorney in such litigation was L. L. James. He soon produced evidence that forms of payment other than cash had not been at all unusual in the early, cash-starved days of the Depression. On occasion, major oil companies had paid lease rentals by giving property owners credit cards limited to the amount of the rental. Property owners in those cases received their rentals in gasoline, lubricants and other products supplied by the companies' service stations.

James filed a motion for a rehearing. It was granted, and after several more years the Court reversed itself and affirmed the judgment of the trial court. Hunt, Continental and Magnolia regained possession of the lease and Hunt was paid $1,500,000 for his part of the oil Gulf had produced.

The case, to most attorneys who watched its progress, was hardly a case at all. It was well established in Texas law that a property owner could waive payment of lease rental if he wanted to. Why then would a Supreme Court hold that he couldn't accept anything but cash? It appeared to be a case in which the law played a minor role.

Even more startling were the large number of suits filed against Barney Skipper and Walter Lechner. Skipper, it will be recalled, had written more than 750 letters to various oil companies and operators begging them to drill his lease acreage. No one had stepped forward. But after Lechner assembled the block and induced John Farrell to drill the Lathrop 1, scores of companies and individuals filed suit to secure a share of the Skipper–Lechner leases, saying that Skipper had promised them shares in his letters. None of them prevailed, but the suits were a nuisance to Skipper and Lechner, both of whom were making fortunes from the leases.

In one of his wry moments, Judge Brown commented from the bench that "nothing, not even the facts, can settle a lawsuit as quickly and as thoroughly as a dry hole." The judge had been hearing a case in which several persons had each claimed title to a small tract of land. Midway through the hearing, word came to the courtroom that a dry hole had been drilled on the tract. Immediately each of the litigants remembered that he had business elsewhere. As he prepared to hear the next case, Judge Brown said to Attorney Lacy, "They flushed out of here like a covey of quail, didn't they?"

Many East Texas lawsuits were holdups, pure and simple. An operator with producing wells might be sued by a promoter who hoped to profit by his nuisance value. In many cases, the operator would prefer to make a quick cash settlement rather than to have his production tied up for as long as a year or more while the case was waiting to be tried.

Most of the land suits developed because it was difficult to prove ownership of hundreds of East Texas farms. Before the boom one farmer might trade several acres to a neighbor for a cow or a horse, but neither would make the trip to the courthouse to have the transaction recorded. Some farms had changed ownership as many as six times with none of the transactions being recorded. Few areas had been properly surveyed. Some of Hunt's leases had open ends: that is, three boundary lines had been staked out but the fourth had not. And a surprising number of small plots had not been surveyed at all. They belonged to no one. With such conditions, and with the prize so great, the thousands of lawsuits were inevitable.

Edwin Lacy was the dean of East Texas title lawyers, and one of the most successful. He won twenty-two lawsuits before he lost one. And he tried the first title suit in the field. With no experience in what to charge for such work, Lacy billed his successful client for $1,000. A few months later, with the boom gaining strength, an attorney winning such a case would have asked for a $20,000 fee and most likely a share of the royalty.

Lacy and other lawyers occasionally accepted lease acreage in payment for their services. Foremost among these was Big Fish. After a while, Fischer had so many oil wells that he often confided to friends that he had been happy, finally, to see the hot-oil business restrained and proration become effective. "It was getting to where every time I worked for a hot-oil client I was working against my own best interests," he explained candidly.

Fischer starred as both attorney and defendant in a series of dramatic, oftimes humorous, trials. Fischer referred to the trials as the "Gabe McIlroy Follies." The case had its roots in ante-bellum East Texas. Gabe McIlroy, a landowner near Tyler, had numbered among his slaves a woman named

Delia. After the war McIlroy freed his slaves and offered some land to Delia for her own. She accepted, and took the name of her former owner.

There was no evidence that Delia had ever married, but she bore a large number of children, all of whom she gave the McIlroy surname. Among them was one Gabe McIlroy, named in honor of the white landowner. When Delia died, the land was divided among her children. Gabe McIlroy's share of the acreage lay in Rusk County.

Gabe McIlroy had left East Texas in 1902 or 1903, leaving behind a daughter—named Delia after her grandmother—and a son, Ambus Jack. They had been living on the land when the boom came. They obtained title to the land by swearing that their father Gabe had been gone for more than seven years, that he could not be located and that he therefore had to be presumed dead. They foolishly leased the land for oil exploration at a dollar an acre—and within a few days had disposed of their royalty interest for a small sum.

The land changed hands several times in the next few years. In 1935, Fischer became the owner. Meanwhile, Delia and Ambus Jack McIlroy had acquired a better idea of the value of their former property. And Fischer began spending many courtroom hours defending his title to the land against a seemingly endless parade of elderly Negroes claiming to be the long-lost Gabe McIlroy. If one were to turn out to be the real Gabe McIlroy, his existence would nullify his children's folly.

Among the claimants—positively identified as their father by Delia and Ambus Jack McIlroy—was a man named Charles Alexander. He was represented by a trio of out-of-state attorneys. Fischer cross-examined him in the trial in Judge Brown's court.

Fischer said, "Now Gabe, when you used to be in this country, you were a little, black, short Negro and you walked like a bear. How come you've changed to a long, tall, yellow Negro?"

The witness replied quickly, "Well, when I left here I took sick. I got some medicine. It was powerful medicine. After I got through taking it, I just growed like I am now, and my color changed on me."

Fischer smiled at him with some admiration. "Now Gabe, when you left here you couldn't read or write. Can you read and write now?"

"No, I sure can't."

"Have you been married since you left here?"

"Oh, no! I was always thinking about my children back here." He smiled lovingly at Delia and Ambus Jack. He was still smiling when Fischer asked the bailiff to bring in his star witness, a woman who had been waiting in the courthouse corridor. She entered the courtroom.

"Do you know this lady?" Fischer asked the claimant.

The claimant hesitated. Fischer bore in. "This is your wife, ain't she . . . your former wife? You lived up here in Daingerfield with her for years, didn't you, and had six or seven children by her? And your real name's Charles Alexander, ain't it?"

The claimant shook his head in despair. "I told those lawyers I never could get by with this lie," he said.

Judge Brown said, "All right, Mr. Sheriff. Put him in jail for a little perjury. And I want to see his attorneys in my chambers right away."

At one time four men, unknown to each other, appeared in Judge Brown's court, each claiming to be Gabe McIlroy. At Fischer's request, Judge Brown consolidated the cases. "Unless all four of you are frauds, only one of you can be the real Gabe McIlroy," Judge Brown said dryly to the claimants. Fischer easily won this case by getting the claimants so confused in their fabrications that the jury was out only ten minutes before giving him the verdict.

The last challenge came in federal court in Tyler. The

claimant was produced by some Oklahoma lawyers who specialized in locating lost Indians who owned oil land. Apparently they had done their homework. At the time of his disappearance in 1902 or 1903, McIlroy had been accustomed to wearing an ancient Prince Albert coat and a tall silk hat. And he almost always carried a Bible under his arm.

Now the claimant appeared in court: an aged Negro wearing an ancient Prince Albert coat and tall silk hat, and carrying a Bible under his arm. And he "walked like a bear" every time he thought of it. When he forgot to shamble, Fischer would say aloud, "He ain't walking like old Gabe now, is he?"

More than fifty witnesses—some of them white—testified that to the best of their knowledge the claimant was Gabe McIlroy. Fischer produced only one witness, a woman.

"Do you know this man sitting there who says he's Gabe McIlroy?" Fischer asked her.

"I know him."

"Who is he?"

"Harry Johnson, my old husband."

"When did you last see this husband of yours?"

"About a month ago."

"Where?"

"He came to the house where I work. That man sitting next to him, that lawyer, was with him."

"Were you surprised to see your husband again after all this time?"

"I sure was. I told him, 'Lord-a-mercy, I thought you was dead years ago!' And he then he says no, that he ain't dead and I ought to come on to the courthouse here and swear he's Gabe McIlroy. He said if he was Gabe McIlroy he'd get some land with oil and lots of money and he'd give me some of the money."

"Did you believe him?"

"I sure didn't! I told him that old Gabe McIlroy had been dead for years. I told him that white men were putting him

up to all this, and they'd take all the money and not give him
any of it. And I told him that even if they did give him some,
he wouldn't give me any of it because he never had given
me any even when we was married."

"How did he take all of this?" Fischer asked.

"Well, he got mighty mad at me. He stomped out of the
house."

"When did you marry Harry Johnson?"

"In 1890."

"How long did you live with him?"

"I'd say until about 1900—except for the two years he was
in the penitentiary."

One of the Oklahoma lawyers jumped to his feet, crying
"Inadmissable! Hearsay! Did you see him in the penitentiary?"

"No," replied the witness before the judge could inter-
vene, "but it ain't no hearsay. He wrote me a letter from the
pen and said he was in there for breaking into a store in
Palestine!"

The Big Fish produced Harry Johnson's prison record, and
the final assault on the Fischer land title collapsed.

The largest, most confusing legal side show in East Texas,
however, was the Virginia case. Its key figure was Paul Hart,
an oil operator described by an attorney of the day as a man
"who had a faculty for uncovering what might be called
'latent values in mixed-up situations.' " Hart had a friend,
Harvey Lewis, who had organized the Lewis Oil Corporation. The corporation owned little or no oil, but Lewis was
constantly promoting. While so doing, he acquired a large
block of stock in the Virginia Company, primarily so he
could obtain the names of stockholders to add to his "sucker
list."

The Virginia Company was organized by Gaines B.
Turner and named for his young daughter. Shortly after
World War I, Turner acquired a number of leases in various
sections of Texas, including some in Gregg County near

Longview. He was a good promoter and sold stock in the company in almost every state of the Union. For a while, he was also a successful oilman. He drilled producing wells in West Texas, but the Gregg County leases he ignored. Like most other oilmen of the period, he grew to consider East Texas barren of oil.

In the late 1920's, the Virginia Company fell on hard times. It was then that Lewis acquired the large block of stock for very little money. In 1930 the Virginia Company went through bankruptcy proceedings in Fort Worth. Its few assets were liquidated for the benefit of creditors. The Gregg County leases were overlooked during the bankruptcy and liquidation, however, because it was assumed that all leases acquired by Turner had expired. That they had not expired became evident when the boom commenced. Several landowners had to turn down enticing lease offers because they had leased their lands to Turner and the Virginia Company for twenty years!

Meanwhile, the Lewis Oil Corporation also had fallen on evil days. Instead of going through bankruptcy, however, a bondholders' committee took over its affairs with the presumed consent of the company's creditors. This committee maintained the semblance of a corporate existence by the election of officers and other steps that gave it some legality. Because he was the largest stockholder, the committee named Paul Hart president of the corporation.

The assets of the corporation were of dubious value, but there was some hope on the part of the committee that recoveries could be made here and there to justify the efforts made to keep the corporation out of complete bankruptcy. It was at this stage that the Lathrop 1 was brought in near Longview. Hart immediately employed a prominent Fort Worth attorney, J. F. O'Brian, to protect and recover the rights of the Lewis Oil Corporation as the major stockholder in the Virginia Company.

The thousands of Virginia Company stockholders had

long ago kissed their investments goodbye. Some had thrown away their certificates and several even had papered their walls with them. But with or without certificates, the stockholders—and creditors—flocked to Texas to get a slice of the pie. Dozens of lawyers swarmed in from all points of the compass.

With the bankruptcy reopened, the lawsuits began. One of the first hearings revealed the possibility that a large amount of Virginia Company stock was overissued, and the field of controversy was extended. By the time that question was resolved—the stock was not overissued—some cases had made their way to the State Supreme Court and one had reached the United States Supreme Court.

Some who held claims sold them for cash. Thus there were substitute stockholders and creditors who deliberately bought into lawsuits in hopes of making a large profit when the lawsuits were finally settled. Among those selling claims were Paul Hart and the Lewis Oil Corporation.

A writer of the day observed: "These lawyers who have flocked into these proceedings are so contentious that it is rather to be doubted whether the creditors in the long run will obtain anything like their expectations, not because the lawyers are unduly voracious but because the questions presented are so complex and, in many cases, unprecedented."

But the creditors of both the Virginia Company and Lewis Oil Corporation did recover most, if not all, of their claims. And those who could prove they were stockholders got their slice of the pie, which turned out to be near the center of the northern sector of the field.

When all was untangled, it was reported that Gaines B. Turner and those associated with him in organizing the Virginia Company had not kept a single share of stock!

chapter twenty

TRAGEDY

Violent death and injury were commonplace in the great field, and one of the most terrible disasters in American history occurred in its very heart.

The boom had hardly started when nine men perished at a Sinclair drill site on the Cole farm near Gladewater. The gusher blew in, gas was somehow ignited, and the well became an inferno. The men burned to death in the first moments of the burst of flame.

A short time later a boomer, his wife and two children died in their tent near Kilgore when oil from a broken pipe-line migrated to their campfire and set the area ablaze. Backfires from automobiles and trucks in gas-filled hollows caused explosions that killed and maimed. Charred metal hulks of tanker-trucks rested on roadsides like burned-out tanks on a battlefield.

While fire was the danger most feared by the workers in the field, day-to-day labor was fraught with peril. In their haste to reach the Woodbine, operators pushed their crews and their equipment. Boilers exploded; derricks toppled; machinery flew to pieces. State and federal governments had not yet established sensible safety regulations—and they likely would have been ignored anyway. Sanitation facilities were almost nonexistent and sickness was rife. Unions had not yet become a power in the American economy. Workers, however, seldom complained; when a man was offered a job during the Depression he didn't inquire about working conditions.

But the frenzy of the boom had waned, and East Texas had achieved a semblance of normalcy when 294 lives were snuffed out in a single explosion that rocked the field from one end to the other.

It will be recalled that at the boom's beginning Tom C. Patten drilled a well—and built his tower—in the only street of the hamlet of London, a community of thirty families. In a short time the hamlet was engulfed by derricks. A second community, called New London, was established nearby to adjoin a highway which ran to Kilgore. And the derricks moved on New London.

The village boasted less than 2,000 inhabitants, but almost 1,000 school-age youngsters lived in the New London Consolidated School District which by 1937 maintained a million-dollar campus. The school district claimed to be one of the wealthiest in the United States; it drew royalty payments from a great number of oil wells on its property, some on the campus itself.

At 3:10 P.M. on March 18, 1937, the main building on the campus was literally blown apart by an explosion that left 280 children and 14 adults dead in the ruin and rubble. So many children were trapped in the wreckage that first estimates of the dead were as high as 600.

Governor James V. Allred, who had succeeded Ma Ferguson, declared martial law and sent in troops to hold back hordes of sightseers who were interfering with rescue operations. Doctors and embalmers were flown to the site from cities such as Shreveport and San Antonio. An identification bureau was set up in the Overton city hall. Oilmen and other rescue workers labored for days before all the children—living and dead—were freed.

With 294 dead there was hardly a family in New London that had not lost a child, a niece or nephew. While messages of sympathy came to the stricken area from around the world, cries for investigations resounded in the state legislature and in Washington. The state sent a military court of inquiry into the area. The U.S. Bureau of Mines began an investigation, and so did soil experts from the Department of Agriculture.

The military court of inquiry decided no one was to blame for the tragedy. The two federal investigations revealed that the explosion was caused by the ignition of a large volume of gas and air that had accumulated in an inadequately ventilated space beneath the building. The gas had escaped through lines or fittings, but investigators were not able to determine if the accumulation was due to slow leakage over a long period of time or to a large break in a gas line just before the blast. The explosion had been touched off by a spark generated by the closing of an electric switch.

The citizenry did not agree with the military court of inquiry that no one was to blame for the disaster. They placed the responsibility squarely on the school district superintendent, W. C. Shaw, on the members of the school board, and on the Parade Gasoline Company. Several months before the blast, Shaw had discontinued the use of gas supplied by the United Gas Company, a public utility, and had "tapped" a pipeline owned by Parade, one which ran near the school. The school system had been paying United about $3,000 a year for gas and service. Gas from the Parade line cost nothing. Shaw's janitors had "tapped" the Parade line and had laid a connecting line into the school building.

Shaw had acted with the knowledge and consent of a

majority of the members of the school board. It had seemed foolish to pay $3,000 a year for gas from United when it could be obtained free from Parade. After all, several churches had "tapped" the line, and some homes and businesses were also connected with it.

Parade was not a public utility as was United, and therefore it could not sell gas directly to consumers. Shaw said he had discussed the line-tapping with one or more Parade officials. He implied that he had received at least tacit approval.

For weeks after the explosion, news reports referred to the gas in the Parade line as "raw" or "wet" gas, the kind that rose from the Woodbine dissolved in the oil. At the surface the oil and gas were separated. Oilmen called the gas "casinghead gas" because it came up the well's casing to the wellhead. Casinghead gas was extremely volatile, containing, as it did, so much liquid. Nevertheless, casinghead gas was widely used to heat and cook in homes and buildings near oil fields. Most of it, however, was burned away in flares; thousands of these tall torches illuminated East Texas by night. Many homes had gas lights in their yards.

The gas in the Parade line, however, was "residue" gas. The liquids, chiefly gasoline, had been "wrung out" of it. To Parade it was waste gas, and the line that Shaw had tapped carried it back to the field for disposal.

Shaw had lost a son and several nephews in the explosion, but he received little sympathy. He was forced to resign. And some thirty damage suits were filed by bereaved parents, naming Shaw, the school board members and Parade as the chief defendants.

On June 24, 1938, the venerable Judge Brown found in favor of all defendants in a representative suit. He had weighed his verdict for several months, the weary jurist said, and had come to the conclusion that "the defendants cannot be held responsible."

Meanwhile, the legislature had enacted a statute prohibiting the use of casinghead gas in Texas schools, and another providing that a substance giving gas a distinctive odor be injected in all gas sold to the public. (It is this odor one refers to when he says he smells gas.) Building standards to promote the safety of school facilities were promulgated. Rules for rigid inspection of school facilities were adopted.

Many native East Texans who had experienced no loss in the explosion nevertheless were traumatized by it. These people felt that the lives of the innocent had been claimed in payment of spiritual debts piled high during the most sinful and reckless periods of the boom. Corruption and lawlessness had gone unpunished; now the blast had exacted the toll. They had drifted away from God as they had waxed fat on the boom; now He had slaughtered the lambs in revenge . . . and as a lesson.

Others bought and sold picture postcards of the disaster scene. A larger, more expensive school building was erected within sight of a hilltop cemetery where most of the victims were buried.

Twenty-four years after the explosion, suppressed memories were revived when a robbery suspect blurted out to Oklahoma City police that he had caused the New London school disaster. The suspect was William Estel Benson, then forty-one. He said he had been a student at the school. He had loosened a gas line in the school basement as revenge against a school official who had reprimanded him for smoking, he said. A relative told police that Benson had confessed his responsibility for the explosion several times within a narrow family circle.

Excitement ran high in East Texas and across the state. Benson was well-remembered in New London. Former schoolmates said that on the day of the blast, they had seen him on the school grounds. But no one had seen him loosen a gas line, police said.

A lie detector test indicated that he had not caused the explosion. Further investigation revealed that Benson had been confined in a mental institution prior to his arrest as a robbery suspect. Police also learned that he had an impressive record as a "compulsive confessor" to crimes he had not committed. Some thirty-six hours after his arrest, Benson finally repudiated his story. The investigation was terminated.

chapter twenty-one

SALT WATER AND THE BIG INCH

The East Texas oil field was one of the country's most valuable assets, but at the time it was needed most it stood in imminent danger of being flooded by salt water from the ancient sea that had formed it. An oil pool which yields 100,000,000 barrels of crude in its lifetime is considered a good oil field; the Black Giant, through 1941, had produced 1,702,915,000 barrels of crude. It also had produced millions upon millions of barrels of salt water; by the end of 1941 it was producing more than 300,000 barrels of water a day! Saltier than the Atlantic or Pacific, the water was polluting rivers and ruining pastures and farmlands. The production of so much water was also lowering the pressure in the reservoir: the original pressure in the reservoir stood at 1,620 pounds per square inch, but by 1941 pressure had dropped to 1,020.71 pounds per square inch—dangerously close to the 750-pound mark at which the natural flow of crude would cease.

This was the situation when the United States became totally committed in World War II. War planners said Texas oil fields were expected to supply 80 percent of the oil necessary for all Allied forces, and the Black Giant would have to provide a major portion of the percentage. Harold Ickes, now chief of the Petroleum Administration for War, declared: "This is a war of machines and of ships and airplanes powered by oil. In short, this is an oil war. The side which, by interrupting the flow of petroleum products to the enemy, and which, at the same time, can supply its own tanks, its mechanized guns, its fighting ships and its airplanes with gasoline, lubricants and fuel oils of the proper kind, at the time required, and in the right places, is the side which will eventually win this world-wide conflict."

But could the Black Giant deliver what was needed, what was expected, when it was almost drowning in its own salt water, when reservoir pressure was dropping at an alarming rate? Could the great field's operators band together in a common cause against a common enemy? The past record held no bright promise.

Salt water had shown up in the field within six months after Joiner completed the Daisy Bradford 3. A Magnolia well on the field's western edge—an 800-barrel-a-day producer—began making salt water with the oil several weeks after it was brought in. A few days later it was producing 720 barrels of water for every 40 barrels of crude. That was the beginning. Well after well on the western boundary began producing salt water with its oil.

The pattern was clear: Withdrawal of a barrel of oil from the reservoir prompted the movement of one barrel of water eastward; too rapid withdrawal of oil in large quantities caused an imbalance in the reservoir mechanism, and geologists and engineers warned that ten barrels of water would be produced for every barrel and a half of oil before the pool was depleted. As was usual in the early years of the boom, the scientists had found few listeners to bad tidings.

Oilmen had been beset with salt water problems since the drilling of the massive Spindletop salt dome near Beaumont in 1901. Salt water from the field contaminated fresh water supplies in the immediate area. Operators were forced to provide safe storage for the salt water during the irrigation season for the area's rice crop. Attempts to return the salt water to the cap rock were unsuccessful.

In 1916 salt water from wells in the Batson, Sour Lake and Saratoga fields in the Big Thicket area of Hardin County threatened the fresh water supply of practically all towns and communities along the lower Neches River. The legislature enacted a measure in 1917 that allowed formation of the Hardin County Salt Water Company, an organization of responsible operators who dug reservoirs and constructed gathering systems to handle the salt water. Successful cooperative efforts then followed in other fields.

But just as the Black Giant contained an ocean of oil, it also contained an ocean of salt water, the disposal of which could not be accomplished by conventional means. As salt water production increased in East Texas, many proposals were offered to rid the field of it. Most were impractical or economically infeasible, and some were simply bizarre. The East Texas Anti-Pollution Committee, an organization of oilmen, reportedly examined them all.

It was calculated that a million barrels of salt water daily could be pumped to the Gulf of Mexico 200 miles away via a 72-inch pipeline constructed of redwood. Cost of building such a line was estimated at $18,000,000—not including expenses for surveying, right of way, and damage to property along the path of the line. Cost of a 48-inch cast-iron line over the same distance was estimated at $30,000,000. Both lines would have to be fed by an extensive $10,000,000 gathering system laid throughout the field. The cost was considered prohibitive.

It was suggested that the salt water be piped into the Sabine River, which cut across the field on its way to the Gulf. Landowners in both Texas and Louisiana protested loudly, claiming their lands and fresh water wells would be destroyed. The idea was abandoned, but many operators surreptitiously dumped their salt water into the Sabine by the hundreds of thousands of barrels. A canal to the Gulf was proposed and forgotten.

Meanwhile, production of salt water increased. Most operators separated the water from the oil at the surface—as they did the gas—then ran the water into what were called "Hope To God Pits." These pits were simply shallow lakes whose sides were built up with a sand and clay mixture. The operators hoped to God that the weak dikes would break or be washed away by rainfall. Their hopes were most often fulfilled, and deserts were created when the salt killed vegetation over many acres.

In 1935 the East Texas Anti-Pollution Committee decided to try a scheme that had been proposed frequently since the salt water had made its first appearance: Return the water to the Woodbine below the oil-producing horizon. If this were done effectively, it would not only dispose of the water but should also maintain or even increase the reservoir pressure. A test site was selected at an abandoned well near Arp, about a mile west of the field's productive boundary. Storage facilities were constructed at the site.

On September 6, 1936, the tests got under way and continued for sixty-one days. During that period, 381,500 barrels of water were run back into the reservoir. Then the Woodbine refused to accept the water. After extensive studies, chemists discovered that exposure to the open air or contact with metal pipe produced bacteria in the salt water. The bacteria was plugging up the space between the sand grains on the "face" of the formation through which the water was to flow.

It was eventually determined that chlorination eliminated the bacteria. Chemists then recommended that in addition to chlorination the "face" of the formation be kept as receptive as possible by backwashing or producing water from the injection well from time to time.

The committee had established that the method would work, but there was no rush to adopt it. It was not until 1938 that the first injection well was put in operation in the field.

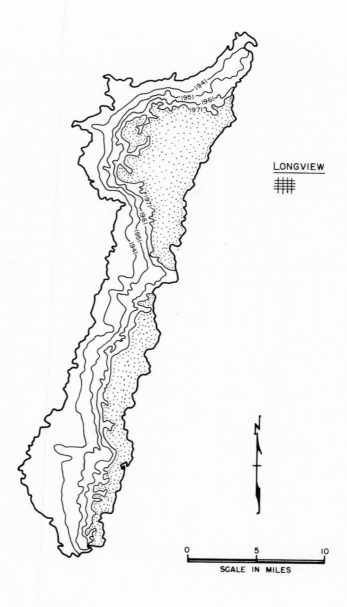

LONGVIEW

SCALE IN MILES
0   5   10

Approximate water-encroachment lines as of 1941, 1951, 1961 and 1971. The shading represents remaining oil-bearing areas as of January 1, 1972.

It disposed of 223,953 barrels of water that year, and in 1939 ten wells—owned by major companies—returned 2,250,761 barrels of water to the reservoir.

But also in 1939 enraged landowners filed a number of injunction suits against operators they claimed were destroying the Neches–Angelina watershed by dumping the salt water in the Angelina River. The suits dragged on, and during this time the first warnings came from Washington that the field might soon be pressed into national service. The federal government made it clear that the salt water problem had to be solved on a field-wide basis, preferably under local control. However, if this could not or would not be done quickly and effectively, the federal government, in the best interests of national preparedness, might have to step in. It was suggested that if a federal agency took charge, the government would not hesitate to shut down all wells producing salt water.

Some of the majors continued to drill injection wells, but they said they could not afford to share them with others. And the smaller companies and operators maintained that they could not drill injection wells or install expensive treating facilities because the volume of water they produced did not justify the expense.

It was finally left to Railroad Commissioner Ernest Thompson to dissolve the inertia. By now Thompson was the best informed and most effective regulatory officer in his field in the oil states. While maintaining his seat on the commission, Thompson had twice run for governor and twice had been soundly defeated. This thwarting of his ambition apparently had not soured him. Indeed, it appeared that he had determined to forget his dreams and dedicate his life to the job at hand. On March 29, 1940, the commission issued an order which provided that the allowable production of any well in the East Texas field that had been converted into a salt injection well could be transferred to the other oil wells on the same lease. Because there would be no loss of oil allowable, Thompson said this was an incentive for operators to convert producing oil wells into injection wells. By the end of the year there were thirty-one injection wells in the field returning 50,000 barrels of salt water a day to the reservoir.

Thompson let it be known that the March order was only a step, that a concerted effort was needed, and that the commission stood ready to aid such an effort. His offer was accepted. At a commission hearing on July 29, 1941, Joe Zeppa of the Delta Drilling Company proposed that he be permitted to drill injection wells to dispose of salt water produced by others. Then Zeppa went further. He proposed that a well owner who returned to the reservoir all of the water produced with the oil from the well be granted an additional oil allowable of one barrel a day. The proposal had the support of several smaller operators in the field, notably Bryan W. Payne, J. H. Edwards and C. R. Starnes. The commission agreed to study the proposals.

On November 20, 1941, the commission issued an order granting a bonus allowable of one barrel of oil per day for each fifty barrels of water returned to the Woodbine. The bonus oil was to be produced from all or any of the wells on the same lease where the salt water was produced. This order was soon amended to allow the bonus to be transferred to wells on other leases owned by the same operator. But the amendment order granted even more. It permitted operators to shut down wells producing more than a hundred barrels of water per day, with the allowable to be produced from another well or wells.

Thompson had concluded that the entire field, and the owners of properties in all parts of the field, would benefit if all or most of the water was returned to the Woodbine. Therefore he listened with an attentive ear to a plan developed by Payne, Zeppa and other operators. It involved creation of a nonprofit corporation that would gather and dispose of all the salt water produced in the field. All operators, large and small, would help bear the cost. Participation would be made possible—and attractive—through the bonus barrel program; the amount of the bonus would be fixed so that the value of bonus oil would be about the same as the cost of salt water injection.

The commission members told the operators to put the plan into operation. Meetings were held, hurdles leaped, and on January 20, 1942, the articles incorporating the East Texas Salt Water Disposal Company were approved by the Texas Secretary of State. Bryan Payne was elected president and Joe Zeppa, the man who had started the ball rolling, was named a director. The company had been started as a $25,-000 corporation. On January 28 the capital stock was increased to $2,000,000, an amount the shareholders considered sufficient to inaugurate a field-wide program. By July 1 the new issue had been oversubscribed.

Some 250 operators in the field—from corporate giants to small independents—had purchased shares. The shares were widely distributed with no one company or individual owning a disproportionate share.

The program was undoubtedly the first ever to be accepted with almost universal approval in the East Texas field. On October 1 the company injected its first barrel of salt water into the reservoir. It marked the beginning of the field's salvation.

It was called the Big Inch. It was the largest oil pipeline ever laid, twenty-four inches in diameter, and it ran from the edge of the East Texas field for some 1,400 miles to the great refining complexes in the New York and Philadelphia areas. It was a wartime necessity, and every day of its wartime service it delivered almost 300,000 barrels of crude to the eastern seaboard.

Another line, the Little Inch, began at a point near Beaumont, Texas, then joined the Big Inch near Little Rock, Arkansas, and paralleled it eastward. It was twenty inches in diameter, and every day of its wartime service it delivered almost 200,000 barrels of aviation gasoline, motor gasoline for tanks and other military vehicles, and other refined products.

Had these lines not been constructed so speedily to deliver such a large amount of fuel, the Allied war effort would have been severely hampered. It took six days for a tanker carrying 100,000 barrels of crude or refined products to sail from a Gulf of Mexico port to New York or Philadelphia—and few were available for such service. German submarines sank U.S. tankers in the Gulf of Mexico, in the Caribbean, off Cape Hatteras, and almost within sight of the Statue of Liberty. The tanker fleet that could be mustered was needed to transport fuel from the eastern seaboard to the war zones.

When the war had begun in Europe in 1939, Railroad Commissioner Thompson had suggested that such pipelines be built; he was cried down as an alarmist. A few months earlier, Glenn McCarthy, a Texas oilman whose brilliance was overshadowed by his flamboyant life style, had suggested the same thing in a letter to President Roosevelt. In July, 1940—between the fall of France and the Battle of Britain—Secretary of the Interior Harold Ickes reported to President Roosevelt that the building of a crude oil pipeline from Texas to the East might prove "absolutely necessary" in the event of war.

If no one in the White House was listening to these suggestions, thoughtful leaders in the oil industry were. In May of 1941, eight senior executives of seven major companies met in New York to discuss the possibility of building emergency pipelines. They concluded that a twenty-four-inch pipeline from Texas to the East could and should be constructed. They appointed W. Alton Jones, president of Cities Service, their chairman, organized National Defense Pipelines, Inc., and earmarked some $80,000,000 as working capital.

Though by that time the German armies were on the road to Moscow, the group's application for priority for steel for the project was turned down. They tried again in September, and again they were refused. Still they continued studying and planning.

The Japanese assault on Pearl Harbor and the subsequent tanker sinkings resolved all doubts in the minds of military and political leaders. Construction of the Big Inch was authorized.

As chief of the Petroleum Administration for War, Ickes called on a bitter foe to head the project—J. R. Parten of Woodley Petroleum Company. Parten had been in the first wave of oilmen to operate in the East Texas field. He had fought shoulder-to-shoulder with Commissioner Thompson to deny Ickes the role he had sought as czar of a federalized oil industry. But Ickes knew Parten was one of the most respected men in the industry, tough, persuasive and capable of rallying support from every segment. He asked Parten to take the job of Director of Transportation for PAW, and Parten took it.

Parten called a meeting of presidents of the eleven largest oil companies operating on the eastern seaboard, including the eight men who had tried earlier to get the pipeline project under way. Parten told them that the government preferred that industry build and operate the line. However, if for any reason the industry was unable to do so, the government would build the line provided that industry would cooperate in a nonprofit corporation to run the project.

At a meeting the next day the presidents reported they were in unanimous agreement that the government should build and operate the line, that each of them would take responsibility as a director of the nonprofit corporation, and that their companies would supply the talent to get the job done. Jones, the Cities Service president, was named chairman of the corporation, War Emergency Pipelines, Inc.

Parten went to the War Production Board to get a steel allocation for the pipe. Steel was needed elsewhere, he was told. He cajoled, he threatened, and he got the first delivery of pipe in July 1942. Meanwhile, the talent supplied by the companies had surveyed the route for the line and had obtained the right of way.

Plans called for the Big Inch to reach Norris City, Illinois, by winter. Railroad tankers would move the oil to the east from that point until the line was completed the following summer. Southwestern railroads protested that eastern railroads would gain a windfall by moving the oil from Norris City to the unloading points. Start the line at Norris City, they said. Let us ship the crude from Longview to Norris City. Midwest pipeline companies suddenly saw the Big Inch as a threat. If the war ends quickly, they said, the huge line can bring gas to our markets at a price we can't meet. W. Lee O'Daniel, junior Senator from Texas, feared that the line would be used to suck all of the gas out of Texas once the war was over.

Hearing after hearing was held—and Parten kept right on with his project. "I don't care what happens *after* the war," he told his critics. Ickes supported him throughout.

Construction superintendent of the vast undertaking was Burt Hull of Texaco, considered the best pipeliner in the business. On August 3, 1942, with the German armies posed before Stalingrad and the Nile Delta, the first joint of the Big Inch was laid. More than 15,000 men worked under Hull's direction. They would take the giant line over eight mountain ridges and under thirty rivers and hundreds of lesser streams. They collapsed from sunstroke, suffered in sub-zero weather, faced the hazards of cave-ins, but they carried on —ten hours a day, seven days a week. Floods tore out the line and they rebuilt it. Machinery was swept away or crashed down mountainsides, and they worked with picks and shovels until the machinery could be replaced.

The first crude from East Texas arrived at Norris City on February 13, 1943, and moved eastward by rail a week later. The following July 19— eleven months and sixteen days after the first joint was laid—the Big Inch was completed at Phoenixville, Pennsylvania. From that point the New York Branch, a twenty-inch line some eighty-five miles long, was

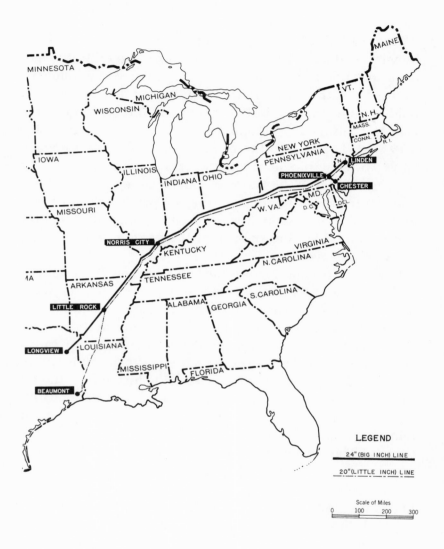

Routes of Big and Little Inch pipelines.

laid to Linden, New Jersey, where distribution lines radiated to the refineries and tank farms at Bayonne, Bayway, Carteret, Tremley Point and Perth Amboy. The Philadelphia Branch extended some twenty-three miles to Chester Junction, with distribution lines reaching refineries at Marcus Hook, Point Breeze and Girard Point.

East Texas crude reached the Phoenixville terminal on August 14 and was delivered to the refineries on the Philadelphia Branch that same day. A week later crude was flowing into the Linden terminal.

Work on the Little Inch had begun on April 27, 1943. The first gasoline arrived at the Linden terminal on March 2, 1944.

In June 1945, with victory in Europe in hand, *World Petroleum Magazine* summed up the pipelines' contribution to the war effort: "Allied invasion of enemy territory in Europe would not have been possible, nor could it have been sustained, without the aid of the 'Big Inch' and 'Little Inch' lines. Constructed in record time, under wartime limitations, the two lines have delivered to the Eastern Seaboard over 316,000,000 barrels of crude oil and refined products."

In the beginning, East Texas supplied all of the crude for the Big Inch. Later it supplied only two-thirds because the crude—so rich in gasoline content—was required in other areas.

During the war years of 1942 through 1945, the Black Giant produced 520,418,000 barrels of oil—as much as five good fields put together could produce in their lifetimes.

As it provided the material for construction of the two pipelines, the War Production Board allotted materials for construction of the East Texas Salt Water Disposal Company's gathering and disposal systems. One was no good without the other.

During the war years the company returned more than

200,000,000 barrels of water to the reservoir—and the reservoir pressure remained almost constant. The company was able to return to the Woodbine about 98 percent of the water it collected. But during the postwar years the company could not have kept pace with salt water production had not the Texas Railroad Commission again come to the rescue. At the beginning of the program the commission had permitted the transfer of oil allowable from any well that was producing at least a hundred barrels of water per day. The transfer, however, had to be to a lease owned by the well's operator. In 1947 the commission ruled that such a well's allowable could be transferred to another operator's lease with the original operator being compensated through a system devised by the commission's staff.

This move choked off a mounting salt water production. Each year saw several hundred water wells closed in, and by 1957 more than 3,000 wells capable of producing a million barrels of water a day had been capped. And by that time the company had injected 1,736,288,451 barrels into the reservoir. Five years later, on its twentieth anniversary, the company reported it had returned the staggering total of 2,424,582,195 barrels of water to the Woodbine. Reservoir pressure had risen to 1,073 pounds per square inch.

The Black Giant appeared healthy. Through 1961 it produced three and a half *billion* barrels of high-grade oil. In 1962 it would produce 43,847,159 barrels more. But also in that year, when stories of its gaudy past were falling fainter on the ear, the most shocking oil scandal since Teapot Dome exploded around its head.

chapter twenty-two

THE SLANT-HOLERS

On the morning of April 17, 1961, Shell crewmen were sent out to rework a well that had been producing for twenty-five years. When they opened the well's valve, fresh drilling mud—not crude as they had expected—issued from the well. The drilling mud surged forth in a familiar rhythm, the strokes of a reciprocating pump. The perplexed crewmen scanned the horizon. The nearest working crew on a derrick was more than half a mile away.

The crewmen approached the derrick. A driller and roughnecks were busy drilling a new well. The Shell crewmen noted the rhythmic workings of the pump, which was sending drilling mud down into the borehole. Later, when the mud pump was shut off at the drill site, the mud quit flowing from the old Shell well. The Shell men watched the drilling crew pull 4,320 feet of drill pipe from the borehole, considerably more pipe than was needed to reach the Woodbine sand.

There was only one conclusion to be reached: The operator was drilling a slant hole from his lease to steal oil from
beneath Shell's lease; his drill bit had accidentally struck the
casing in the old Shell well and punctured it, allowing the
drilling mud to escape through the Shell well. The operator's lease was in an area where salt water encroachment
had all but drowned out the crude in the Woodbine. The
Shell lease was good for at least seven more years of production.

The operator at first denied any wrongdoing, but later
admitted that he had "got caught." Nevertheless, the Texas
Railroad Commission granted the operator a permit which
allowed him to "straighten out" his borehole. The permit
was granted at a public hearing in which Shell representatives asked that an inclination survey be run on the well as
it was being drilled, with Shell engineers in attendance. The
second part of the request indicated that Shell's representatives had little confidence in commission employees to conduct a proper survey. The incident attracted little if any
attention. To save face, the operator drilled a straight hole
to the Woodbine. He found so little oil as to make the venture unprofitable. It was his only punishment.

A few months after the Shell incident, on July 25, 1961, J.
D. Matthews, a private investigator for Humble, shot and
killed Leonard Dorsey, an oil-field worker, in a Henderson
motel room. Matthews was tried on a murder charge. He
pleaded self-defense and was acquitted. Evidence was produced that Matthews was investigating oil thefts in East
Texas for Humble, and that Dorsey had offered to sell him
information about illegal drilling operations.

Only the facts of the slaying attracted any attention.
There was little discussion about Matthews' employment
and Dorsey's offer to sell information.

But on April 10, 1962—a year after the Shell incident—Jim
Drummond, Houston bureau chief of the Chicago-based *Oil
Daily*, called the Railroad Commission in Austin. He spoke

with William J. Murray, the commission chairman. "There are reports that the commission has evidence that some operators in East Texas are stealing oil from their neighbors by means of slanted wells," Drummond told Murray. "Are the reports correct?"

Murray did not answer immediately. Drummond waited patiently; he already owned some indisputable facts. Finally Murray said, "The reports are correct. A certain amount of noncompliance with regulations on drilling deviation has been uncovered . . ."

Regulations held that no well in the East Texas field should deviate from the vertical by more than three degrees. Murray said thirty-one wells had been severed from their pipeline connections in the past year because they deviated beyond the prescribed limit. Most of them were along the eastern flank of the field. Since the reservoir's water drive pushed the oil from west to east, the eastern edge of the pool was the area where the last barrels of oil from the Woodbine eventually would be produced. Operators had drilled slanted holes from barren acreage beyond the limits of the field back into the productive Woodbine. They had hoped to stay in business for the remaining life of the field.

Drummond's story in *The Oil Daily* the next day lifted the scab from a sore that had been festering in East Texas for fourteen years. A series of investigations followed. Before they had run their course it was concluded that some 400 deviated wells—almost all of them in East Texas—had siphoned off a billion dollars in oil from the rightful owners! Civil penalties suits were filed against 150 firms and individuals, and 163 persons were named in criminal indictments. Oil companies filed scores of damage suits against oil pirates and scores of settlements were made without court action.

The thieves were not jailbirds or get-rich-quick operators. Most of them were pillars of the East Texas community. Many were known for their philanthropies and regular church attendance. Some were established oilmen or respected landowners; others were lawyers, industrialists and businessmen. Two were judges. They had been aided and abetted by crooked drillers, engineers, roughnecks and corrupt Railroad Commission personnel.

In Texas, where headline scandals sometimes appear to be the rule rather than the exception, this one was the tawdriest of all. Yet, when all the furor died, not a single oil thief had spent a single day in prison.

The Railroad Commission, though derided and censured during the investigations, emerged intact as an operating entity. Cries for a new commission to supervise the state's oil and gas production went unheeded as they had gone unheeded in the thirties. A Gregg County grand jury, which handed down indictments against a number of prominent East Texans, declared, "It is unthinkable that this condition could have existed for nearly twenty years without the Railroad Commission's taking action to correct illegal practices."

At the time of the scandal the commission was composed of William Murray as chairman, Ernest Thompson, who was gravely ill, and Ben Ramsey, a newcomer. Murray therefore bore the brunt of the criticism aimed at the commission. And like Thompson, he had been a member of the commission in 1948 when reports of slant-hole drilling first began circulating in the field. Even then it was understood that the thieves were drilling from beyond the eastern limits of the field westward into the prolific reservoir. It was obvious, too, that the commissioners at least suspected it, for they ordered that a "straight-hole clause" be included in all permits issued under exceptions to the spacing rule for operations along the eastern boundary. That was in August of 1948. On April 1, 1949, the commission put into force Rule 54, which required operators to obtain special permits and fulfill specific requirements to drill directional wells. Among the requirements were regular progress reports in the form of directional surveys. These surveys indicated not only the degree of deviation from the vertical but ultimately showed whether the well bottomed on the operator's property. The thieves obviously ignored Rule 54.

On April 19, 1950, Edwin G. Stanley, the commission's Kilgore district engineer, addressed a highly significant letter to his superiors in Austin. Stanley was the son of Captain E. N. Stanley, Thompson's enforcer in the early days of the boom. In his letter he said, "There is a dire need for additional restrictions on drilling of wells in the East Texas field. This office has received various rumors as to the drilling of directional deviated wells; although none has been substantiated, there are certain indications that there has been and will be wells directionally drilled in areas previously proven dry and completed as producing wells."

He seriously questioned the "accuracy of the inclination surveys filed with the Commission." Inclination surveys were designed to show the degree of deviation from the vertical, but—in contrast to directional surveys—gave no clue as to the ultimate bottom of the deviated well. The inclination survey simply showed the degree of deviation without giving the direction in which the borehole was being drilled.

Stanley ended his letter with several recommendations: Directional surveys should be conducted by operators under the strict supervision of commission personnel; only in cases where the deviation was within the three-degree limit should wells be granted their oil allowables. The Kilgore office should be authorized to conduct inclination surveys "on all questionable wells," and the commission should authorize the Kilgore district supervisor to "require a directional survey on wells he deems advisable."

On May 2, 1950, a memorandum signed by Murray and Thompson went out to all operators in the field. It contained the Stanley recommendations almost verbatim. Apparently all three recommendations were quite often honored in the breach by crooked operators and corrupt commission employees. Ironically, when the scandal broke in 1962, Stanley —who had since left commission employ to become an oil operator—was caught up in the investigations and charged.

From 1948 until the slant-hole artist pierced the Shell well in April of 1961, the commission took no action against the thieves though almost 400 crooked wells were drilled in the East Texas field and a score more in other fields about the state. However, in the year between the Shell incident and Jim Drummond's telephone call to Murray, thirty-one deviated wells had been shut off, Murray said. No other action had been taken against the operators.

But with the publication of Drummond's story, the storm broke. A legislative committee opened public hearings, airing the whole sordid mess. Grand juries in Gregg, Rusk and Upshur Counties commenced investigations. Oil companies that had been victimized filed huge damage suits. Attorney General Will Wilson sent fifty Texas Rangers into the field to assist in the search for slanted wells. There had been rumors of impending violence, and Wilson told reporters, "We don't anticipate any violence, but we're ready—just in case." And the federal government began a sort of investigation, too. It was just like old times.

And, just as in boom days, the miscreants walked among their fellow citizens with heads high, crying that "the big boys are out to get us." Some eighty of them banded together and with an audacity which would have been preposterous outside East Texas, offered to "mediate" the scandal. Some of them served up a twisted but persuasive logic. "Look," they said in effect to their victims. "When we took a barrel of crude from your leases on the eastern flank of the field, the water drive pushed one in to take its place. You didn't lose anything. The guys on the west side of the field were the ones who were hurt."

Just as in boom days, the state had few funds available for a detailed search for deviated wells. As they had done in the boom days, some majors and large independents donated money to help the state. Delays such as this gave some operators an opportunity to plug their slanted wells with cement so they could not be examined by conventional means. Others switched identification markers on their wells so that investigators examined straight-hole wells. Commissioner Murray admitted at a press conference that commission agents had been gulled into inspecting the same well several times, each time with a different identification marker. Still, the number of slant wells located mounted. By October of 1962 more than 150 had been found, and others were discovered with every passing week.

At one point the state ordered that all wells on a lease containing a slant well be shut down. This brought frowns to the faces of East Texas businessmen and politicians. The thieves had been paying taxes on the stolen oil as well as on oil legitimately produced. Rumors were born that the East Texas economy might suffer a crippling blow. Cessation of illegal drilling had caused some unemployment. Supply companies had been dealing with the thieves, and their business had slacked off. Now, with the closing down of all wells on a lease with a slant-hole well, local politicians and school administrators began to fret about reductions in tax revenues.

The first two criminal cases were brought to trial in East Texas. The defendants were acquitted. The state therefore abandoned that course and filed civil penalties charges against the accused, with trials to be held in Austin. The penalty for illegally producing oil was $1,000 for each day of production. The state softened the blow by declaring that an operator would be accountable for only thirty days of illegal production from each slanted well.

Reluctant to face Austin juries, some of the accused rushed to Austin to settle up. Some left the state. Others declared themselves bankrupt. Some went to trial.

The state had filed civil penalties suits asking for more

than $26,000,000 from the defendants. When the books were finally closed, the state had received $1,117,000 with no prospect of getting more. But the pirates had been hurt financially, nonetheless. Straight wells on leases containing slant-hole wells were shut down for two years, preventing production of 21,000 barrels of oil daily at an annual loss of $23,000,000. The slanted wells were plugged forever.

Most companies that had filed damage suits had to settle before trial for pittances. Thieves who had been living high on the hog pled poverty. Their bank accounts could not be located. Companies seldom fared better when their suits went to trial. Texaco, for example, sued one operator for $2,000,000 and was awarded $16,000 by an East Texas jury. Animosity toward the "big boys," born in the thirties, obviously had not died away.

The thieves had been shrewd enough to classify their slant wells as "marginal producers," which under commission rules were wells incapable of producing more than 20 barrels of oil per day. Marginal wells were not subject to proration and thus could produce to capacity every day of a month. Other wells—good wells—were limited to producing only eight days a month. Thus a marginal well producing 19 barrels daily would produce 570 barrels each month, while a good well making 30 barrels a day on an eight-day producing schedule could produce only 240 barrels. The thieves, in many instances, were making more money per well than their victims!

Commission employees took an operator's word that his well or wells were marginal. "No one," declared the Gregg County grand jury, "has inspected or supervised the marginal-well fiasco." It was customary for a slant-holer to obtain a lease on the field's eastern flank, one out of the producing area but near enough to reach it with a deviated well. By

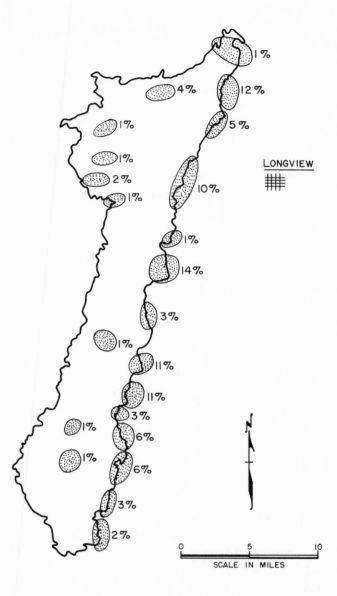

Shaded areas represent locations where slant-hole drilling occurred. The figures denote the percent of slant-hole wells drilled in each particular area.

declaring that he had found only marginal production, he reduced suspicion while insuring a greater profit. Some slant-holers placed dummy wells near the deviated wells. The dummies—hardly more than pipes stuck in the ground —were linked to the slanted wells by underground pipelines and were also presented as marginal producers.

Some of the wells had been "kicked" to an angle of 60 degrees. The basic tool of the slant-holers was the "whipstock," a twelve-foot length of pipe whose cross-section presented the picture of a triangle. In a directional drilling operation, the drill bit entered the pipe and traveled along the triangle's hypotenuse. This forced the bit from its vertical path. The whipstock was often used more times than one in reaching the Woodbine. It was complex work requiring more than ordinary skills. It called for engineers, expert drillers and moonless nights. It helped to have the consent of commission employees, five of whom were fired or resigned during the early days of the scandal. One was accused of taking more than $150,000 in bribes.

The Railroad Commission was under fire early and late until the scandal ran its course. The thieves and their supporters charged the commission with taking the side of the "big boys," while others voiced wonderment that the commission had not taken action against the thieves at an earlier date. Another critic was Attorney General Will Wilson, who told reporters shortly after the scandal erupted: "There is no doubt someone has been closing their eyes or this couldn't have gone as far as it has."

Wilson seemed to run away with the state's investigation. He was almost constantly in the headlines or on the radio or television. This appeared to annoy Commission Chairman Murray. He delivered a major address entitled "Scandal or Survival" before the Dallas meeting of the Texas Mid-Continent Oil and Gas Association on October 9, 1962, in which he implied that Wilson was more interested in the former than the latter.

"Your speaker, as an engineer, has always disliked publicity-seeking titles," Murray said, "but lately the Railroad Commission and its chairman have been sinking into oblivion without ever being mentioned in the press, on radio or on television." He paused, then added: "You well know this is a facetious statement, and the Railroad Commission yearns for less publicity, fewer interviews and more opportunity to proceed with its important work."

Wilson apparently felt it necessary to reply; he issued a statement that he was not seeking membership on the commission.

Six months after the speech, Murray resigned from the commission midst charges of conflict of interest. It was disclosed that his personal income from oil-related activities during his commission tenure had exceeded $1,700,-000. None of the activities were connected with the slant-hole scandal, but the scandal brought them into public view.

The legislative hearings produced some fireworks when four members of the Federal Petroleum Board office in Kilgore stood on federal privilege and refused to testify. Richard S. Allemann, an attorney for the Department of the Interior, gave the legislators a rather unconvincing explanation of why the four should remain silent. "It is essential to the success of our investigative and prospective efforts that undue publicity be avoided," Allemann said. "Of even greater concern is the maintenance of the traditional constitutional lines of authority which separate the powers of the several states and the powers of the federal government."

It was generally accepted that federal agents were hunting for any violations of the Connally Hot Oil Act, which forbade shipping of contraband oil in interstate commerce. Allemann said the slant-hole problem was strictly a Texas headache, that the Connally Hot Oil Act "was designed to implement and assist your enforcement of your laws."

Federal reluctance to make full disclosure of any facts available to the Kilgore office of the Petroleum Board led to considerable speculation, most of it political. Some of it was dispelled when the legislators called for testimony from two former chairmen of the Federal Petroleum Board. James R. Lewis, who had been chairman from 1948 to 1953, said he was fired for pressing an investigation of slant drilling after being told to abandon it "by a man sent down from Washington." His Washington superiors, Lewis said, felt the slant-hole situation was an affair for the state to deal with. As for the extent of slant-hole drilling, Lewis said, "I don't see how anybody close to the operation there [East Texas] would not have been put on notice."

Nelson Puett, board chief from 1955 to 1960, said his staff did not rely on or cooperate with Railroad Commission agents "because the information would get back to the oil operators."

Tarnish wears off quickly under the hot Texas sun. Many of the oil pirates retained their civic stature during the scandal. Perhaps East Texans thought the public exposure punishment enough. Perhaps the magnitude of the thefts engendered more envy than distaste. Certainly there was an undercurrent of admiration for men with the temerity to steal from the large oil companies.

And certainly the incentive was there for acquisitive men of weak moral fiber. The federal government had frozen the price of crude at $1.25 a barrel during the war. So great was the postwar demand for automobiles and fuels to run them that East Texas crude sold for $2.65 a barrel by 1948, when slant-hole drilling began. The price rose year by year, reaching $3.20 a barrel by 1957, and stood at $3.10 a barrel in 1962 when the scandal erupted.

A description of the thieves which lingers was offered by George Wear, a Continental attorney, in a speech before the Houston Kiwanis Club. "Who are these illegal operators that have continued and expanded their illegal operations over an extended period of time?" Wear asked. "They are often men who attend their civic club meeting on Wednesday, play golf on Saturday, attend the church of their choice on Sunday and appear as leaders in their civic, social and religious organizations.

"Without the normal identification of a crook, they have been beyond suspicion. With wealth, community and political prestige, they have been able to fool the gullible, compromise the unwary and bribe as they encountered their kind. There may be among their lot the true independent operator who makes the finding of oil his life's work; but as a rule they could not be so classified because they are takers, not finders. They will believe that they have acquired a license to steal, will scream of injustice and persecution as the practices are brought to a halt. They will seek sympathy, probe for further weaknesses and having once found such weaknesses will expect continued success.

"They will not only use all the means and accumulated wealth to avoid the consequences of their acts, but they will expect to continue in business as illegal operators."

Slant-hole drilling, Wear said, meant "trespass, theft, bribery and a willful, intentional violation of laws and regulations. It means an intentional underground trespass on the property of another to steal. It means the filing of false reports and the making of false affidavits. It means taking advantage of the trust of others, corruption and bribery."

Wear's speech implied that the Black Giant's wealth would always be a target of the unscrupulous. The implication reminds one of a bitter joke, a play on words, that was heard wherever oilmen gathered during the scandal: A driller leaving a drilling rig waves goodbye to his replacement, saying, "See ya later, Deviator."

chapter twenty-three

THE LAST BOOM

After more than forty years, the Black Giant is still the largest operating oil field on the North American continent, with two billion barrels of top-grade crude yet remaining in the reservoir. Four billion barrels already have been taken from its Woodbine blanket. In each year of its existence it has produced more oil than an unusually good oil field produces in a lifetime.

The boom-and-bust days are over. The days of reckless drilling are past. Twice the Black Giant went to its knees— once from over-drilling, then from salt water encroachment —before men and companies learned their lesson. Oil conservation was born in East Texas, and the rules and regulations to affect it were established there through trial and error. It has become a primary fact of life in the oil business.

The old days meant wanton loss of reservoirs from the too-rapid withdrawals by men and companies bent on sudden fortunes. Now slow and sensible drainage allows the oil sands the full measure of their lives in which to yield the full measure of their bounty. Where in the past, oilmen were fortunate to recover as much as 40 percent of the crude from a reservoir, they now look forward to recovering as much as 80 or even 90 percent. And the country has benefited, for a nation that moves on oil needs every drop from every reservoir her oilmen can discover. Had the lesson not been learned, the United States probably would have become dependent on foreign oil before World War II.

Perhaps the great field wrought changes in the industry's men as well as the industry. The early-day American oilman, at home or abroad, always was a more colorful figure than the American cowboy, if for no other reason than that he played for higher stakes. He still is colorful: the lonesome wildcatter thirsts on the high plain and slogs through cruel swamps; the great companies plumb the earth beneath the sea and send their bits through the tundra in the frozen northlands. But their thoughts and attitudes have changed to a great degree since the rough-and-ready days of the Last Boom. The die-hard executives of big companies who felt that the only answer to tough competition was the freeze-out, and the unscrupulous independents who sought wealth at any cost to others, began fading from the scene. A sense of moral and public responsibility began to permeate the industry. Its critics may not believe this, but it is true.

Oilmen have a saying, "Only the drill finds oil." This is not to belittle their scientists. With all of their knowledge and devices, geologists and geophysicists can only find the structures in which oil is likely to accumulate; they can only point to the likely place to drill. The drill must be spun into the earth by someone with the courage and finances to back the scientists' judgment.

Geologists call the East Texas field a stratigraphic trap. The Black Giant gave no surface or buried indications of its presence to the scientists of the 1930's—and no stratigraphic trap of like consequence has been discovered in this country since the Black Giant was found by the drill. There may be many more Black Giants beneath American soil; some highly respected geologists think there are. They believe that intensive efforts to locate them should be made to assure an energy-short America of a future oil supply; that lacking scientific evidence of their existence, the industry should apply the drill on the vast acreage which has not yet been touched. And they believe that scientists should make a concentrated study to learn more about the elusive stratigraphic traps. America, they point out, may not forever be permitted to *freely* import oil from foreign lands.

These geologists are in effect calling for more wildcatters, whether they be major companies or individuals with a hunch and the nerve to play it.

Columbus Marion Joiner is dead. So are Daisy Bradford, Doc A. D. Lloyd and Ed Laster. So are Walter and Leota Tucker. For seventeen years after Joiner brought in the Daisy Bradford 3 the old wildcatter sought another great field in vain. Morris County, Titus County, Pecos County—all felt his tread. His last exploration was in Comanche County. His "nose for oil" had not picked up the proper scent. He remained in vigorous good health until a few months before his death in Dallas in 1947. He was almost eighty-seven years old.

The Daisy Bradford 3 is still producing. Almost every morning of the year a pumper flips a switch and the old well coughs up a barrel or two. If you leave the highway at the right time of day and drive for a mile along a crooked road through tall green trees, the pumper may fill a small bottle with good crude from the old well and let you take it home with you. Few people take the trouble, yet 10,000 found their way to the remote spot in 1930 to witness the birth of the Black Giant.

The pumper also may point you to a stone marker the State of Texas set up near the well in 1936. Push aside the underbrush and you see the words: "Discovery Well (Joiner #3 Bradford) brought in September 3, 1930, as the first producer in the largest oil field in the world." It is a cold marker, and the date is wrong.

Miss Daisy never saw it. She died as the boom reached its frenzied height. She also was denied the sight of the schools, libraries, parks and paved streets Joiner had promised his drilling would create. The pre-boom hamlets became fine and beautiful cities, proud of their heritage but equally proud of the civic accomplishments of their people.

H. L. Hunt still owns the Daisy Bradford 3. For him the old well produces memories that keep him young. The fortune Hunt built in East Texas served as the foundation for one much larger, for he could no more stop hunting for oil than could Joiner—and he seemed to find it as often as not. He is regarded as one of the richest individuals in the world, and is best known to the multitudes for his wealth and for what the majority consider his political eccentricities. But in the oil patches of the world, from East Texas to the Middle East, the people who know him best call him "one hell of an oilman." And that's the appellation he likes best.

Hunt remembers. He remembers with delight and affection the people and the action: the beauty of Dad Joiner's language, Miss Daisy's gentle manner, Walter Tucker's unbounded enthusiasm for any project, the tart tongue and generous spirit of Leota. And he remembers Doc Lloyd, "that strange and brilliant man." Who knows the paths Lloyd wandered from the East Texas field to his deathbed in a Chicago hotel in 1941? Like Joiner, he was almost eighty-seven. Hunt shook his head when told of Lloyd's death, and murmured, "What great dreams thundered in that man's skull."

J. Malcom Crim, the man responsible for the second great well in the field, is dead, and so is Ed Bateman, the promoter

who put the deal together. And the well, the Lou Della Crim
1, was abandoned and plugged on April 13, 1961. Before it
gasped its last breath the well had produced 250,000 barrels
of crude. Crim, the conservative businessman who believed
a fortuneteller, lived a long and fruitful life as one of Kilgore's most beloved and respected citizens. Bateman
retired to a ranch in West Texas—and oil was discovered
beneath his pastures!

Barney Skipper, the Longview Prophet, is dead. He was
honored as few prophets are honored in their hometowns.
The Lathrop 1 was the third great well in the field, but
Skipper is remembered in Longview as the man who pro-
phesied an "ocean of oil" a dozen years before Dad Joiner
began his poetic pitch in Rusk County.

Walter Lechner, the only man who ever listened to Barney Skipper, is still alive—and at eighty-one still working
every day. Lechner took Skipper's dream and turned it into
reality. The Lathrop 1 has never been put on a pump; the
crude flows from the well as readily as it did forty years ago.

Judge R. T. Brown, the Sage of East Texas, died in 1952.
He had spent his entire life in Rusk County. Prestigious law
firms had tried to lure him from the bench; Judge Brown
always had replied: "As long as the people want me on the
bench, that's where I'll stay." Most of the boom-time lawyers
who practiced in his courtroom are dead, including F. W.
Fischer, the Big Fish. Fischer once said of Judge Brown, "He
arrives at a just decision more often than any judge I ever
saw simply by finding out what the hell is going on." Few
ever argued with that opinion.

Ernest Thompson is dead. It was Lieutenant General
Thompson at his passing; in 1948 he was appointed commander of the Texas National Guard and promoted to lieutenant general. But Thompson had earned another title
which the oil fraternity had bestowed on him—the Father
of Petroleum Conservation. He had served continuously on
the Railroad Commission from 1932 until his retirement in

1965, the longest any man had ever served in public office in Texas. He died the next year. With each passing year of his tenure on the commission he had grown in stature; he was generally regarded as the most knowledgeable and able man in the oil and gas regulatory field.

Many who knew the great oil field from the beginning are alive, and some are still active. Lone Wolf Gonzaullas, at eighty-one, is chief security officer for a large Dallas apartment complex. He still wears his revolver, and he's still the handsomest man on his beat. E. O. Buck, the commission engineer who made the first official study of the Black Giant, lives in Houston. So do L. T. (Slim) Barrow, who retired as Humble's chairman, and Francis X. Bostick, the paleontologist-geologist who advised Ed Bateman and urged the promoter to keep drilling on the Lou Della Crim 1. Another Houstonian is J. Howard Marshall, who supervised the federal government's legal efforts against the hot-oil runners. Tom Kelliher, the former FBI agent who put the hot-oil runners out of business, now lives in retirement in Willis, Texas. W. W. Zingery, the mapmaker who initiated the receivership suit against Joiner, is a Houstonian and still in business.

Ed Zilkey, the former baseball player who helped Ed Bateman drill the Lou Della Crim 1, lives in Fort Worth. Dry Hole Byrd, who loaned Dad Joiner equipment to complete the Daisy Bradford 3, lives in Dallas. E. A. Wendlandt, the early-day Humble geologist, lives in Tyler. So do Henry Conway, the Amerada scout on the discovery well, and Joe Zeppa, who helped conceive the East Texas Salt Water Disposal District. Ronald Reese, the first oil scout ever to visit the Daisy Bradford 3, lives in Jackson, Mississippi. J. R. Parten, the East Texas oilman who pushed construction of the Big Inch, lives in Madisonville, Texas.

The East Texas field is bright in their memories. Each knows that he helped make industrial history there. They were a part of the Last Boom.

In the summer of 1971, a friend was visiting H. L. Hunt in the oilman's Dallas office. It seemed a shame, the friend said, that Hunt had never met Walter Lechner. "After all, he was a pioneer in the East Texas field—and you've been neighbors for thirty-five years or more. His office is just a block away."

"When will you see him again?" Hunt asked.

"I've got a luncheon appointment with him."

"Invite him up here," Hunt said promptly. "We'll have lunch right here."

Now Walter Lechner for years has lunched daily at the Imperial Club in the Baker Hotel. He enjoys a shot of Old Fitzgerald or Jack Daniels before attacking a prime rib of beef and assorted vegetables.

But on this day he sat with Hunt and the friend at a coffee table in Hunt's office. He had a glass of juice, broth heated on an office hot plate, sliced onions and carrots, and peanut butter with bread or brown crackers.

He ate with relish as he and Hunt talked of old times, of people they had known, of places they had been. They parted, reluctant to end the conversation.

Later Lechner was asked if he had enjoyed the food. His thick eyebrows jumped. Truth struggled with loyalty to a new acquaintance, Hunt, and to the sharing of revived memories. Truth lost. "Hell," snorted Lechner, "it's a diet I ought to have been on myself for twenty years!"

an acknowledgment

More than two hundred persons were interviewed in the preparation of *The Last Boom*, and to them we extend our thanks for sharing their memories with us. Some of them, old friends, died while the work was in progress. Like the living, they had wanted to see the Black Giant's story between book covers. Perhaps some young man or woman will read it in their stead, and find therein a bit of the joy the oldtimers knew when they struggled with the Black Giant.

The authors are grateful for the aid provided by the publishers, editors and reporters of *The Oil Daily*, the *Oil and Gas Journal*, the Kilgore *News Herald*, the Henderson *Daily News*, the Longview *Morning Journal* and *Daily News*, the Tyler *Courier-Times*, the Houston *Chronicle*, the Houston *Post*, the Fort Worth *Star-Telegram* and the Dallas *News*. The Black Giant provided a steady flow of news for their pages.

A prime source of material was the library of the Energy Research and Education Foundation, Houston. Other sources were *This Fascinating Oil Business*, by Max W. Ball, Douglas Ball, and Daniel S. Turner (Indianapolis: Bobbs-Merrill Company, Inc.); *The Growth of Integrated Oil Companies*, by John G. McLean and Robert W. Haigh (Harvard University Graduate School of Business Administration); *History of Humble Oil and Refining Company*, by Henrietta M. Larson and Kenneth Wiggins Porter (New York: Harper & Brothers); *Some Ran Hot,* by Ruel McDaniel (Dallas: Regional Press); *Three Stars for the Colonel,* by James A. Clark (New York: Random House); *East Texas Oil Parade,* Harry Harter (San Antonio: Naylor Company).

appendix

GEOLOGICAL, TOPOGRAPHICAL AND PETROLIFEROUS SURVEY, PORTION OF RUSK COUNTY, TEXAS, MADE FOR C.M. JOINER BY A.D. LLOYD, GEOLOGIST AND PETROLEUM ENGINEER

TOPOGRAPHY

The Overton Anticline is located in the Juan Ximinez, Isaac G. Parker, R. W. Smith, Geo. W. Guthrie, Bennett Smith, T. J. Moore and M. J. Prue original surveys, or land grants which have been subdivided into small tracts belonging to the present numerous owners, reference to which is hereby made to Rusk County ownership map as compiled by C. M. Joiner.

The area covered by this survey is marked by the presence of high-rounded, steep-sided, sand-covered hills, with narrow streams traversing narrow, sandy valleys. The Overton Anticline is located on a ridge from which the streams flow North into the Sabine River, and South into the Neches River. The smaller streams are fed with springs and seldom become entirely dry. The larger streams are deep and sluggish and have the general appearance of Gulf Coast Bayous. The higher hills are covered with Pine, and the valleys are marked by the presence of Live Oak, Post Oak, Gum, Maple and Willow trees. The streams and forests will furnish ample water and fuel for developing purposes. The area is traversed by hard-surfaced highways with laterals that are generally graded and passable for heavy equipment.

STRATIGRAPHY

The formations deposited in the regions surrounding the Overton Anticline belong to the Wilcox Eocene (Claiborne Group). The formations dip South from the Overton Anticline which is marked by the presence of the Yegua Formation until within twenty-five miles distance from the Overton Anticline; the Cook Mountain and Mount Selman formations form the surface sediments. On the North and East, Cook Mountain is the highest formation encountered within twenty-five miles. On the North and West nothing higher than the Cook Mountain formation was identified during this survey.

The areas covered by Rusk, Cherokee and portions of Smith Counties, Texas, have suffered greater distortion generally than any other similar-sized area in East Texas or Louisiana with the exception of that area covered by the Sabine Uplift in N.W. Louisiana; and there are no other areas of similar magnitude where the position of the strata can be interpreted by engineers with the degree of certainty that is possible in this area. This enables structural conditions to be defined with a degree of certainty quite unusual for Delta Structures. During the time now elapsed since the bringing in of the Humble Oil Co.'s No. 1 Discovery Well at Carey Lake, Jacksonville, Cherokee County, Texas, this regional high has attracted the attention of all the major oil companies as well as the important independent operators, and large blocks of acreage have been leased by the

Roxana, Shell Oil Co., Magnolia Oil Co., Texas Co., and the Humphries Co. The above blocks are all in the counties mentioned, and the Overton Anticline (C.M. Joiner Block for whom this report is made) is located in the center of the above leased blocks. Much of the land lying between the above named blocks has been leased by independent companies who paid unusual prices for same.

FAULTS, FOLDS AND DIPS

The Overton Anticline is located near a Saline Dome that has generally disturbed the strata of the region. This dome is marked on the surface by a Salt Lake, or Marsh, in the N.E. Corner of a 104.75-acre tract now belonging to the Mayfield Co., and by the presence of a Salt Crystal that forms on the bottom of a well dug for water on the N.E. Corner of a one-hundred acre tract belonging to Calvin Young, and by the presence of a Salt Lick on the S.E. Corner of a 74.5-acre tract belonging to J. A. Birdwell. All the above is in the M. J. Prue Headright Survey. The above occurrence of salt in the earth and water forms a triangle with a lateral line of about one mile, and is from two to three miles from the apex of the Overton Structure.

The Overton Anticline, while lying close to a Saline Dome, is not a Saline Dome Structure, but is a Faulted Anticline traversed by an East and West Fault extending in a general direction sixty-eight degrees North by West. This Fault Course is crossed on the highway between Arp and Overton, where it is marked by a well-defined arch, and on the easterly extension it becomes the arch in the city of Henderson, county seat of Rusk County. This fault may be traced for a considerable distance in both East and West Extensions. On the East it can be traced as far as the Shelby County Oil Fields; on the West across Smith and Van Zandt Counties. This anticline is also traversed by a fault extending in a general direction North twenty-two degrees East. The

North Extension of the fault would traverse the Caddo Oil Pool in the region of the Pine Island Saline; the South Extension traverses the Humble Oil Co., Cherokee County Saline where the No. 1 Discovery Well is located.

The red, white, blue and brown stratified sandstones and shales are sufficiently bedded and consolidated to enable clinometrical measuring of the dips that have been recently exposed by the cuts made in grading the highways of this region, and a great number of these dips have thus been exposed at numerous places where they develop arching, doming, and possible faulting, and the structures that may be developed by these exposures justify, in the author's estimation, the present activity displayed in this district. The Aerial Geology of the Overton Anticline is well marked and developed by the well-exposed strata dipping in all the cardinal directions from the apex of the anticline. These dips have enabled the petroleum engineers to place the seismograph in the most advantageous position to receive the concusionare vibrations. Under these favorable conditions thousands of registrations have been made on and around this block by major companies.

ANTICLINE AND DEPTH TO PAY

The Overton Anticline displacement has probably resulted from the intrusion of the Saline Core to the southwest, and the upthrow of the traversing fault is found on the south and east quadrants and the downthrow on the North and East of the intersection of faults.

There is a drag toward the apex in the Conture Lines of the N.W. and S.W. Quadrants, which are the larger quadrants of the structure. There are two outlying small, slightly elevated Dome Areas lying just off the Central Drag of the Conture Lines of these two major quadrants. These two outliers will be important producers. The entire structural condition forms a geometrical arrangement that is exceptionally balanced, which justifies the conclusion that the accumulation of oil and gas will be of unusual importance. The author expects some very large wells to be developed on this structure.

The location of the C. M. Joiner Well No. 1, on the Daisy Bradford Tract near Johnson Creek will spud in the top of the Yegua Formation on the base of the Cook Mountain. Fossil leaves and plants of undoubtable Yegua were obtained at a depth of twenty-five feet from a water well dug on the H. M. Cooper thirty-five acre tract which showed similar fossilization. Salenite Crystals have been found in several water wells and reported as Mica by the land owners. Many of the water wells have water with unusual taste that accompanies the presence of Alum and Gypsum. These salts and fossil remains, and the dips which enable the strata to be traced for great lateral distances substantiates the conclusion that this well will begin in the top of the Yegua formation.

The log of the Rucker Well, drilled on the Artic Wright Land located on the Parmelia Chisum Survey about five miles North by East from Henderson, and eight miles East by North from C. M. Joiner Location No. 1 encounters the top of the Austin Creek Chalk at 1,497 feet. The strata between the Rucker Well, located in a syncline, and the C. M. Joiner Location are sufficiently exposed and continuously connected to correlate between the two wells, and such correlation shows that the Rucker Well cut 175 feet plus what will be cut by the Joiner Well. This also substantiates the dip to the Northeast.

Base of Austin Chalk in Rucker Well	2500	feet
Top of gas sand in Rucker Well	3270	"
Base of Austin Chalk in Humble Cherokee Well	3004	"
Top of Oil Sand in Humble Cherokee Well	3843	"

Which checks the formations in
these wells within 69 feet
Base of the Austin Chalk in the
Rucker Well 2500 "
Top of Limestone in Rucker Well 500 "
Base of Austin Chalk in Humble
Cherokee Well 3004 "
Top of Limestone in Humble
Cherokee Well 3706 "

which correlates the formation and pay sands in the two
wells to within a few feet. The C. M. Joiner Well should
encounter pay in the stratum from which the Rucker Well
produced gas at a depth of 3230 to 3270, at a depth of 3055
after deducting 175 feet of strata cut by the Rucker Well,
which will not be cut at the Joiner Location.

These correlates show the thickness of the sediments are
continuous and uniform over an area from the Humble Oil
Co., Cherokee County Well to the Joiner Location in Rusk
County.

THE PRODUCING OIL SAND

The producing zone in both the Rucker and Humble occurs at the zone of the first limestone and beneath the Eagle
Ford Shales, which zone is classified as the Woodbine Sand.
There is a possibility of picking up a pay in the Nacatosh
Sand at a shallower depth. The Rucker Well picked up a
stray shallow sand at a depth of from 577 to 604 feet. This
pay horizon will be encountered at a depth of 402 to 429
feet. The Trinity Sands should be encountered at a depth of
4200 feet. These producing oil and gas sands in the fields
now developed in the region surrounding the Joiner Well
yield large gushers.

Location Joiner Well No. 1 is made on the Daisy Bradford
Tract in the Juan Ximinez Headright Survey.

index

About the Author

JAMES A. CLARK and MICHEL T. HALBOUTY are coauthors of *Spindletop*, the story of the oil discovery that changed the world.

Both men were on the ground during the great East Texas boom, Halbouty as resident geologist and engineer for a large independent oil company, Clark as a journalist getting the first interview with Dad Joiner when his discovery well came in.

Clark is an author, biographer, historian, and columnist. He knows the practical side of the petroleum industry and has steeped himself in its lore. A wildcatter bucking the odds is a more romantic figure to Clark than a trail boss goading a herd of longhorns to the railhead. In 1972 he received the first American Association of Petroleum Geologists Award for Journalism.

Halbouty is one of the world's eminent petroleum geologists and engineers. He is past president of the American Association of Petroleum Geologists, the world's largest organization of earth scientists. In addition to many other awards, in 1971 he received the coveted DeGolyer Distinguished Service Medal of the Society of Petroleum Engineers of the American Institute of Mining, Metallurgical, and Petroleum Engineers. He is the author of several books and numerous papers on scientific and technical subjects, and is an internationally known lecturer and public speaker.